Mathematics

Problem-Solving Challenges
for Secondary School Students
and Beyond

Problem Solving in Mathematics and Beyond

Series Editor: Dr. Alfred S. Posamentier
Chief Liaison for International Academic Affairs
Professor Emeritus of Mathematics Education
CCNY - City University of New York

Long Island University
1 University Plaza -- M101
Brooklyn, New York 11201

Published

Problem Solving in
Mathematics and Beyond | Volume **04**

Mathematics
Problem-Solving Challenges for Secondary School Students and Beyond

David Linker
Alan Sultan
City University of New York, USA

World Scientific

NEW JERSEY · LONDON · SINGAPORE · BEIJING · SHANGHAI · HONG KONG · TAIPEI · CHENNAI · TOKYO

Published by

World Scientific Publishing Co. Pte. Ltd.

5 Toh Tuck Link, Singapore 596224

USA office: 27 Warren Street, Suite 401-402, Hackensack, NJ 07601

UK office: 57 Shelton Street, Covent Garden, London WC2H 9HE

Library of Congress Cataloging-in-Publication Data

Names: Linker, David (Mathematics teacher) | Sultan, Alan.

Title: Mathematics problem-solving challenges for secondary school students and beyond /
    David Linker & Alan Sultan (City University of New York, USA).

Description: New Jersey : World Scientific, 2016. | Series: Problem solving in
    mathematics and beyond ; v. 4

Identifiers: LCCN 2015036879 | ISBN 9789814730037 (pbk : alk. paper)

Subjects: LCSH: Mathematics--Problems, exercises, etc. |
    Mathematics--Competitions. | Mathematics--Study and teaching (Secondary)

Classification: LCC QA43 .L6584 2016 | DDC 510.76--dc23

LC record available at http://lccn.loc.gov/2015036879

British Library Cataloguing-in-Publication Data

A catalogue record for this book is available from the British Library.

In-house Editors: V. Vishnu Mohan/Tan Rok Ting

Typeset by Stallion Press
Email: enquiries@stallionpress.com

Printed in Singapore

Preface

You are about to embark on a journey in mathematical problem solving where you will encounter some rather unusual and challenging problems many of which originated in the New York City Interscholastic Math League (NYCIML) contests. The problems as well as the solutions will very likely appeal to the beginner as well as to the experienced problem solver as the problems vary in scope and contain many rich ideas. The solutions provided consist of those that were officially published, as well as some of our own, and some by a very talented young man, Danny Zhu who gave his own unique perspective on many of the problems. The problems have been arranged according to what the authors felt are Level 1 (attainable by most students), Level 2 (problems which require tricks or somewhat more thought) and Level 3 (problems which the authors consider more challenging). Of course, what one considers a Level 3 problem, others might consider a Level 1 problem, so these decisions should be taken merely as suggestions.

For the past few years, the NYCIML offered 4 levels of high school contests each year: senior A, senior B, junior, and freshmen-sophomore contests. The problems from the different levels have been combined in this book. Questions of a certain type that frequently occurred were represented by a typical exemplar of this problem type.

This book may be used by teachers who wish to have a resource of many non-traditional problems to challenge the brighter students, as well as for anyone who enjoys solving mathematics problems where the solutions are not always straightforward. This is also an excellent resource for math-team participants who wish to prepare for high school contests, and to enrich math-team coaches preparing to train their students for contests. Thus, this book can appeal to a rather wide audience.

While the authors have tried to give detailed solutions, they have left some of the basic calculations associated with solving equations to the reader. So, if the problem requires solving a quadratic equation, they will often just say "The solution to this quadratic equation is...," feeling confident that the reader can fill in the details of solving the quadratic equation. Also, although certain problems can be easily done with a calculator, the goal is to find a way to solve the problem without a calculator, since calculators are not permitted to be used at the contests. Often problems require quick thinking and excellent calculating skills, since students typically only have 10 minutes per set of two problems.

An appendix is provided with lists of formulas and theorems that can be helpful to solve some of the problems. The lists are not exhaustive, as any person who reads this book is expected to know certain basic facts, like the area of a circle is πr^2, or that $x^2 \cdot x^3 = x^5$. Consequently, these simple facts will not be included throughout the book. Some of the facts included are quite sophisticated, and it would be an interesting challenge for the student, the teacher, or the general reader to try to prove those results (for example, Ptolemy's theorem concerning a cyclic quadrilateral).

To do well on these contest problems, one will have to know a bit more than the standard high school material. The intent here is to expand the reader's knowledge and boundaries of topics that are outside of the ordinary, but not necessarily beyond the level of high school mathematics. For example, knowledge of modular arithmetic makes many of the problems in number theory much easier to solve. Therefore, high school students who participated in these contests often independently broadened their knowledge of mathematics beyond the high school curriculum — laterally, not necessarily vertically. Again, basic facts are provided in the appendix on modular arithmetic and other related topics, but the reader may wish to refer to the internet or any basic book on number theory to become better versed in this, and related results as often alternate solutions are provided using these concepts.

The book has been divided into nine chapters, by subject matter. Of course, many problems can fit in a variety of different areas of mathematics. For example, a geometry problem may use the law of cosines, leading to a quadratic equation; thus, requiring the use of trigonometry and algebra. Or, a probability problem may use geometry in its solution. Thus, categorizing problems by subject requires making a choice as to what is the main thrust of the problem, and what topics it uses. The book begins with problems with arithmetic and logic, and then moves along to algebra, geometry, trigonometry, logarithms, counting problems, number theory, probability, and functional equations. The computational problems in the arithmetic section use basic arithmetic, but sometimes require a bit of deeper insight. For example, a problem might assume a knowledge of the laws of exponents.

The authors are always eager to learn about clever alternate solutions, and therefore would be pleased to hear from you should you have any. David Linker may be contacted at mathprofccny@gmail.com and Alan Sultan may be contacted at asultan956@aol.com.

Alfred S. Posamentier

July, 2015

Acknowledgments

By author, David Linker: I have been collecting math problems since I was in high school. It was always my intent to publish them and share them with others interested in problem solving and contest mathematics. I was encouraged to do this by Dr. Alfred Posamentier, who provided guidance, and helped keep the writing on track, as well as providing fine suggestions along the writing process. As the long-time head coach of the New York City Math Teams, I got to work with some of the most talented and dedicated mathematicians and educators in the field. To acknowledge a few, all of whom taught me something about mathematics and problem solving, the brilliant Larry Zimmerman, the exceptional David Hankin, the extraordinary Oleg Kryzhanovsky, and the incredibly dedicated Richard Geller, who passed far too soon. Some of the students who taught me methods were Aleksandr Khazanov, part of the US Mathematics Olympics Dream Team that scored a perfect score in the Olympics in Hong Kong in 1994 and whose life ended tragically at an early age. Jan Siwanowicz also competed in the Olympics, was a Putnam fellow, and is one of the best mathematicians, best problem solvers, and kindest people I have known. He and others like Oana Pascu, Joel Lewis, and Ming Jack Po have always been there to advance my skills.

I want to thank my wonderful wife Toni, for her encouragement and understanding. I particularly want to thank my co-author, Dr. Alan Sultan. Alan took much of the raw materials we had and put them together in a coherent way to produce, what I hope will be a book enjoyed by many.

By author Alan Sultan: I would like to thank my wife, Ann, for her understanding and encouragement during the construction of this book.

Both authors wish to give special thanks to the publisher, World Scientific and their helpful and knowledgeable staff, including V. Vishnu Mohan, Tan Rok Ting and Rochelle Kronzek, without whom this book would not have been possible. And special thanks to United States Physics Olympic's Gold Medalist Danny Zhu for his diagrams and alternate solutions and for making, what we hoped was a good manuscript, even better.

About the Authors

In high school, author **David Linker** was always interested in mathematics problem solving and contests. He received his BS in mathematics from Queens College of the City of New York, and his MS from the Belfer Graduate School of Science. After college, he began teaching high school mathematics and started a Math Team class that won numerous competitions over the course of ten years. He was chosen to be president of the New York City Interscholastic Mathematics League (NYCIML) and served as president, chief problem editor, and contributing author for 11 years. During this time, he also became the Head Coach of the New York City Math Teams. He chose the 68 best high school mathematicians in New York City and went to the New York State Mathematics League (NYSML) competitions, where the teams were undefeated. He also took the teams to the national American Regions Mathematics League (ARML) competitions, where the teams finished in the top 5. In 2000 he was appointed the director of the Math and Science Center at the City College of New York. He became the director of the Kaplan and Petrie Institutes for the Advancement of Mathematics Education. He was a founder and instructor at the City College Summer Scholars Academy, where he helped train students for mathematical tests and contests, such as the AMC, AIME, and the Mathematical Olympiads. He has been teaching at the City College of New York, specializing in problem solving and technology. He is a contributing author to numerous problem solving and geometry books. He has lectured extensively at Limacon and other venues on problem solving.

Alan Sultan is a professor of mathematics at Queens College of the City University of New York. He is the author or coauthor of about 40 articles; some of which are research articles and some of which are expository in nature. He has also authored or coauthored 5 books besides this one, with the following titles: "Linear Programming (Second Edition)", "A Primer on Real Analysis, The Mathematics That Every Secondary Math Teacher Needs to Know", "Implementing the Common Core Standards Through Problem Solving, Middle School", and "Implementing the Common Core Standards Through Problem Solving, High School". He is also the recipient of several grants. In his spare time, he sings bass in two choruses and performs at several different venues throughout the year.

Contents

Assumptions and Understandings

Notation that we use is standard in mathematics.

$\overline{AB}$ indicates line segment AB.

AB means the length of the segment. For example, if we say compute AB, we mean to find the value of the length of the segment.

$\overleftrightarrow{AB}$ refers to the line.

$\overrightarrow{AB}$ refers to the ray.

$m\angle ABC$ refers to the measure of the angle.

Compute means to find the answer in simplest form. So in a contest, $\frac{2}{4}$ would not be an acceptable answer, but $\frac{1}{2}$ or 0.5 is acceptable. We also required answers to have rational denominators.

Iff means if and only if.

$x \in (0°, 360°)$ means that x is an angle between $0°$ and $360°$, but not $0°$ or $360°$.

$x \in [0°, 360°]$ means that x is an angle between $0°$ and $360°$, and can be $0°$ or $360°$.

This can be written as $\{x \mid 0° \leq x \leq 360°\}$.

Chapter 1

Arithmetic and Logic

Questions

Level 1

1. Four houses each have four floors. On each floor are four apartments, with four doors in each apartment. On each door are four hinges. How many of these hinges are there in the four houses?

2. The average of 50 test scores is 38. One of the scores is 45, while another is 55. If these papers are removed, compute the average of the remaining scores.

3. The average of Jeff's nine test grades is 92. If his highest grade, 96, and his lowest grade are not counted, the average of the other grades is 94. Compute his lowest grade.

4. We have five quarts of a solution which is made of 60% acid and 40% water. If 5 more quarts of water are added to this solution, compute the percentage of water in the new solution.

5. A town is built on a grid system of two-way roads. North–south roads are numbered consecutively as streets (1st Street, 2nd Street, etc.), while east–west roads are numbered consecutively as avenues (1st Avenue, 2nd Avenue, etc.). Billy lives on the corner of 75th Street and 83rd Avenue, while Hope lives on the corner of 92nd Street and 65th Avenue. Compute the minimum number of blocks that Billy can drive to go from his house to Hope's house, assuming he stays on the roads.

6. A clock gains five seconds per hour. Twenty-four hours after the clock is set correctly, it reads 5:58. What is the correct time?

7. Compute the smallest two-digit positive integer that is the sum of the cube of one of its digits and the square of the other digit.

8. In the addition of two four-digit numbers below, each different letter represents a different digit. Compute the value of the sum.

$$
\begin{array}{r}
AABB \\
+BACC \\
\hline
CCBA
\end{array}
$$

9. Compute the number of digits in the number $4^{16} \cdot 5^{20}$.

Level 2

10. A train to Washington, D.C. leaves from New York at noon and travels at a constant rate of 70 miles per hour. Also at noon, a train to New York leaves from Washington, D.C. along an adjacent and parallel track traveling at a constant rate of 40 miles per hour. They will meet each other somewhere along the route. How far away from each other will they be 1 hour before they meet?

11. An urn contains 70 marbles which differ only in color. Of the marbles, 20 each are red, green, and blue. The remaining 10 are either black or white. Without looking, and without replacement, n marbles are removed from the urn. Find the smallest n that ensures that at least 10 marbles of the same color have been removed.

12. How many degrees are there in the angle formed by the hands of an analog clock at 7:20?

13. Compute how many different integers from 100 to 400 inclusive are perfect powers (perfect squares, perfect cubes, etc.).

14. There are 100 students in the sophomore class of Bawne High School. Of them, 43 students take biology, 35 take physics, 42 take chemistry, 12 take physics and chemistry, 15 take biology and physics, 17 take biology and chemistry, and 7 take all three courses. Find how many students take none of the courses.

15. Find the units digit of $1! + 2! + \cdots + 14! + 15!$.

16. Find the 1991st digit to the right of the decimal point in the decimal expansion of $\frac{1}{14}$.

17. In the table below, we begin writing the positive integers in row D and move to the right, writing each consecutive integer until we reach row G, then we continue writing consecutive integers in the next row moving left until we reach row A, and then continue

writing in the next row moving right, and so on. Find the column that contains the number 1998.

A	B	C	D	E	F	G
			1	2	3	4
11	10	9	8	7	6	5
12	13	14	15	16	17	18
	$\cdots$		22	21	20	19

18. A cubic block with 8-inch edges is painted and then cut into 1-inch cubes. Compute the number of cubes that have at least one face painted.

19. How many integers between 1000 and 5000 are perfect squares?

20. A train begins crossing a 450-meter-long bridge at noon, and clears the bridge completely 45 seconds later. A stationary observer notices that the train takes 15 seconds to pass him. Compute the speed of the train in meters per second.

21. The inhabitants of a certain kingdom will always tell the truth when asked when their birthday is, except if they are asked on their actual birthday. Mrs. Larken (one such inhabitant) was asked what her birthday was on both January 5th and January 6th. She replied, "It was yesterday," both times. When is her birthday?

22. Find the sum of all the digits of the integers between 1 and 1000 inclusive.

23. Compute the sum of the positive integral divisors of 448.

24. Find the units digit of $3^{88} \cdot 7^{87}$.

25. Compute the sum of all three-digit positive numbers that have only odd digits.

26. The number 12003000 ends with a set of three consecutive zeros. How many consecutive zeros does 50! end with?

27. Find the units digit of $999^{(999^{(999^{999})})}$.

28. Compute

$$\cfrac{1}{1 + \cfrac{1}{1 + \cfrac{1}{1 + \cfrac{1}{1+1}}}}.$$

29. Scientists believe that the Earth's rotation is gradually slowing down. Suppose that in a billion or so years, the Earth will turn on its axis so slowly that a single day in the future will be 30 of our

present hours long. By how many present hours will a future week (7 future days) be longer than our present week?

30. We are given the following statements:
    (a) Harry is a seeker, or Hermione is smart.
    (b) If Harry is a seeker, then Fluffy is dangerous.
    (c) If Hermione is smart, then Voldemort is evil.
    (d) If Fluffy is not dangerous, then Voldemort is not evil.
    Based on these statements, can we be sure that (a) Fluffy is dangerous, (b) Fluffy is *not* dangerous, or (c) neither?

31. Let n be a four-digit integer whose digits are all distinct. If $9n$ equals the integer obtained by reversing the digits of n, compute n.

Level 3

32. Compute the sum of all possible four-digit positive integers formed by using each prime digit exactly once.

33. Compute the sum of all of the digits of the even integers from 1 to 1000 inclusive.

34. Compute the sum of the digits of all multiples of 25 from 1 to 10^5 inclusive.

35. Compute the units digit of $1^5 + 2^5 + 3^5 + \cdots + 2006^5$.

Answers

Level 1

1. In total, there are 4 houses, 4^2 floors, 4^3 apartments, 4^4 doors, and $4^5 = \boxed{1024}$ hinges.

2. **Solution 1:** The sum of all the scores is $50 \cdot 38 = 1900$. The sum of the remaining scores is $1900 - 55 - 45 = 1800$, so the average of the remaining scores is $\frac{1800}{48} = \boxed{\frac{75}{2}}$.

   Solution 2: The average of the removed scores is 50, and there are 24 times as many remaining scores as removed scores. Thus removing the scores pushes the average away from 50 by $\frac{1}{24}$ of the distance from 38 to 50, which is $\frac{1}{2}$, so the new average is $38 - \frac{1}{2} = \boxed{\frac{75}{2}}$.

3. **Solution 1:** The sum of all 9 grades is $9 \cdot 92 = 828$ and the sum of the remaining 7 grades is $7 \cdot 94 = 658$, so the sum of the removed grades is $828 - 658 = 170$, and the unknown grade is $170 - 96 = \boxed{74}$.
   Solution 2: Since removing 2 grades and leaving 7 increased the average by 2, the average of the removed grades is less than the average of all the grades by $\frac{7}{2} \cdot 2 = 7$, so it is $92 - 7 = 85$. The unknown removed grade is therefore $85 + (85 - 96) = \boxed{74}$.

4. The original 5 quarts contained $0.4 \cdot 5 = 2$ quarts of water. Adding 5 more quarts of water gives 10 liters in total, of which 7 are water. Thus, the percentage of water is $\boxed{70\%}$.

5. In total, Billy must travel at least $92 - 75 = 17$ blocks across streets and $83 - 65 = 18$ blocks across avenues. Simply doing one and then the other leads to a shortest path of length $\boxed{35}$ blocks.

6. In 24 hours, the clock gains $24 \cdot 5 = 120$ seconds, or 2 minutes, so the correct time is $\boxed{5:56}$.

7. The number cannot start with 1, since the only possible sums between 10 and 19 are $1^3 + 3^2 = 10$ and $1^3 + 4^2 = 17$, which do not work. If the number starts with 2, we see $2^3 + 4^2 = \boxed{24}$.

8. Since the last two columns of the addends are identical, but the last two digits of the sum are different, there must be a carry from the units column into the tens, and therefore, also a carry from the tens into the hundreds. Thus, the units column gives $B + C = 10 + A$ and the tens column gives $B + C + 1 = 10 + B$, so $C = 9$ and $B = A + 1$. The hundreds column gives $A + A + 1 = C$ (or $10 + C$, but that would imply $A = C = 9$ (contradicting, $A \neq C$)), so $A = 4$, $B = 5$, and the sum is $\boxed{9954}$.

9. $4^{16} \cdot 5^{20} = 2^{32} \cdot 5^{20} = 2^{12} \cdot 10^{20}$, so the number is $2^{12} = 4096$ followed by 20 zeros, for a total of $\boxed{24}$ digits.

Level 2

10. In the hour before they meet, the first train travels 70 miles and the second 40 miles, so they are $70 + 40 = \boxed{110}$ miles apart at the beginning of that hour.

11. We may have as many as 9 each of red, green, and blue marbles, as well as 10 of the combined black and white marbles (since they might not be all black or all white). However, if we add one more marble, it must be red, green, or blue, and we will have 10 of that color. Thus, the minimum is $9 + 9 + 9 + 10 + 1 = \boxed{38}$.

12. The number of degrees between any two consecutive numerals on a clock is $\frac{360°}{12} = 30°$. At 7:20 (Fig. 1.1), $\frac{1}{3}$ of an hour has passed since 7:00, so the hour hand is $10°$ past the numeral 7. The minute hand is on the numeral 4, so the angle between them is $(7-4)\cdot30°+10° = \boxed{100°}$.

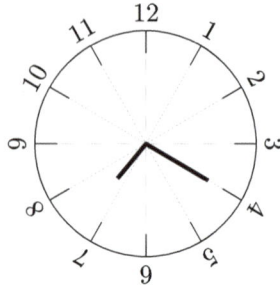

Figure 1.1

13. There are 11 perfect squares (10^2 through 20^2); 3 perfect cubes (5^3 through 7^3); 1 perfect fourth power, which was already counted ($4^4 = 16^2$); 1 perfect fifth power (3^5); 1 perfect seventh power (2^7); and none of any other power. This is a total of $\boxed{16}$ powers.

14. **Solution 1:** By the principle of inclusion–exclusion, the number of students taking at least one course is $43+35+42-12-15-17+7 = 83$, so the number of students taking no courses is $100 - 83 = \boxed{17}$. **Solution 2:** Consider the Venn diagram of the courses shown below (Fig. 1.2). Since there are 7 students taking all the courses and 17 taking biology and chemistry, there must be $17 - 7 = 10$ who are taking only biology and chemistry, but *not* physics. Similarly, there are 8 taking only biology and physics, and 5 taking only physics and chemistry. Now, there are 43 students taking biology, and 10, 7, and 8 taking the possible combinations of biology with at least one other class, so there are $43 - 10 - 7 - 8 = 18$ taking only biology. Similarly, there are 20 taking only chemistry and 15 taking only physics. Finally, the number of students taking at least one course is the sum of the numbers in all the regions of the diagram, which is 83, so the number of students taking no courses is $100 - 83 = \boxed{17}$.

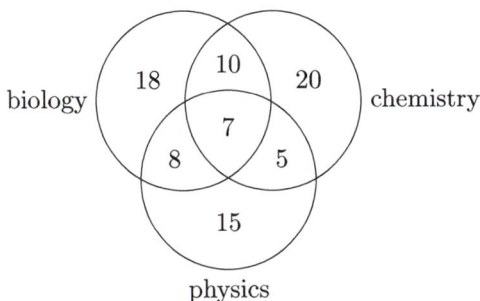

Figure 1.2

15. All the terms from 5! onward are multiples of both 2 and 5, and hence 10, so they all end in 0. So we just need the units digit of $1! + 2! + 3! + 4! = 33$, which is $\boxed{3}$.

16. We have $\frac{1}{14} = 0.0\overline{714285}$, which repeats in cycles of length 6 after the first 0. (You can see this quickly, based on the more commonly-known multiples of $\frac{1}{7}$, since $\frac{1}{14} = \frac{5}{70} = \frac{1}{10}\frac{5}{7} = \frac{1}{10} \cdot 0.\overline{714285}$.) Therefore, we want the 1990th digit, starting from the first 7. Since 1990 yields a remainder of 4 when divided by 6, the answer is the 4th digit in the cycle, which is $\boxed{2}$.

17. **Solution 1:** Every number in column D is 1 more than a multiple of 7. If the number is 1 more than an odd multiple of 7, we are moving to the left in that row as we fill in the numbers, and moving right otherwise. Since 1995 is an odd multiple of 7, we know 1996 is in column D, and we need to only move 2 spaces to the left to find 1998, which gets to row $\boxed{B}$.
    Solution 2: For those readers who are familiar with modular arithmetic, another approach: The pattern of columns repeats every 14 numbers; 2002 is a multiple of 14, so $1998 \equiv -4 \equiv 10 \pmod{14}$, and so 1998 is in the same column as 10, which is $\boxed{B}$.

18. The painted blocks form a 1-inch thick layer around the cube; removing them would leave a 6-inch cube, which would contain 6^3 out of the original 8^3 blocks. Thus, the outer layer contains $8^3 - 6^3 = 512 - 216 = \boxed{296}$ blocks.

19. Since $31^2 < 1000 < 32^2$ and $70^2 < 5000 < 71^2$, the smallest square is 32^2 and the largest is 70^2, so there are $70 - 32 + 1 = \boxed{39}$ squares.

20. The train starts crossing the bridge when its front end reaches the near end of the bridge and it finishes crossing when its back end reaches the far end of the bridge (Fig. 1.3); in the 45 seconds between those moments, it travels its own length plus the length of the bridge. It takes 15 seconds to travel its own length, so it takes 30 seconds to travel the length of the bridge, and its speed is $\frac{450\,\text{m}}{30\,\text{s}} = \boxed{15\,\text{m/s}}$.

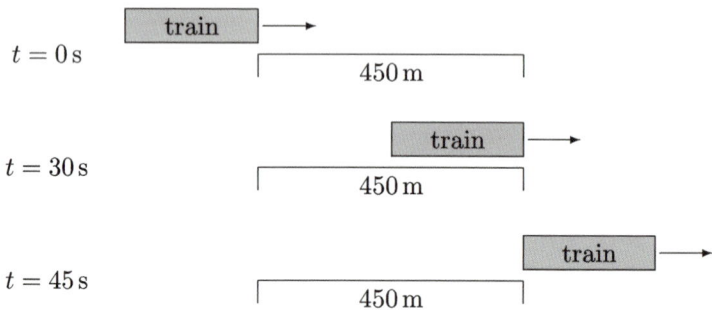

Figure 1.3

21. Since the answers are contradictory, one must be false, so her birthday is one of the days on which she was asked. If her answer from the 5th were true, her birthday would not be one of those days. Thus, her birthday is the $\boxed{5^{\text{th}}}$.

22. We can pad each number with 0s up to three digits without affecting the sum; then, apart from 1000, we have the numbers $000, 001, 002, \ldots, 999$. This list contains 3000 digits in total; by symmetry, each of the 10 digits must appear 300 times. Thus, the sum of all the digits is $300 \cdot (0 + 1 + \cdots + 9) = 300 \cdot 45 = 13500$. Then, 1000 contributes an additional 1 to the sum, for a total of $\boxed{13501}$.

23. **Solution 1:** The prime factorization of 448 is $2^6 \cdot 7^1$, so every factor of it must have the form 2^a or $7 \cdot 2^a$ for $a = 0, 1, \ldots, 6$. Thus, the sum is $(1 + 2 + \cdots + 2^6) + 7 \cdot (1 + 2 + \cdots + 2^6) = (7 + 1) \cdot (2^7 - 1) = \boxed{1016}$.
    Solution 2: The general formula for the sum of divisors gives $\frac{2^{6+1} - 1}{2 - 1} \frac{7^{1+1} - 1}{7 - 1} = 127 \cdot 8 = \boxed{1016}$.

24. The units digits of successive powers of 3 cycle through 3, 9, 7, and 1 in that order. Since 88 is a multiple of 4, the units digit of 3^{88} is 1. Similarly, the units digits of powers of 7 cycle through 7, 9, 3,

and 1, so the units digit of 7^{87} is 3. Thus, the units digit of the product is $1 \cdot 3 = \boxed{3}$.

25. Each digit can be any of the 5 odd digits, so there are $5^3 = 125$ numbers in total. For any such number except 555, you can replace each of its digits d with $10 - d$ to get a different valid number (for example, 153 would become 957). Thus, the average of all such pairs is 555, so the average of all the numbers is 555, so their sum is $125 \cdot 555 = \boxed{69375}$.

26. A zero on the end will be produced by a factor of 2 and a factor of 5. Since there are more factors of 2 in this product than there are factors of 5, we will always be able to match a factor of 5 with a factor of 2 to get a 0 on the end. Therefore, we need only count the factors of 5. There are 10 numbers from 1 to 50 having a factor of 5, and 25 and 50 have two factors of 5, one of which we already counted. So we have two additional factors of 5 for a total of 12 factors of 5. Thus, the number of zeros at the end is $\boxed{12}$.

27. Since an odd number raised to any power is odd, our number is 999^k for some odd k. When a number ending in 9 is raised to an odd power, the units digit is always $\boxed{9}$.

28. **Solution 1:** Working from the bottom up, we have

$$\cfrac{1}{1+\cfrac{1}{1+\cfrac{1}{1+\frac{1}{1+1}}}} = \cfrac{1}{1+\cfrac{1}{1+\frac{1}{1+\frac{1}{2}}}} = \cfrac{1}{1+\cfrac{1}{1+\frac{2}{3}}} = \cfrac{1}{1+\frac{3}{5}} = \boxed{\frac{5}{8}}.$$

Solution 2: For those readers who are familiar with induction: Let $a_1 = 1$ and $a_{n+1} = \frac{1}{1+a_n}$. By induction, $a_n = \frac{F_n}{F_{n+1}}$, where $F_1 = F_2 = 1$ and $F_{n+2} = F_{n+1} + F_n$. Then the number we want is a_5; we have $F_5 = 5$ and $F_6 = 8$, so $a_5 = \boxed{\frac{5}{8}}$.

29. A present week is $7 \cdot 24$ hours long and a future week will be $7 \cdot 30$ hours long, so the difference is $7 \cdot (30 - 24) = \boxed{42}$.

30. Refer to the individual parts of the statements by "*seeker*", "*smart*", "*evil*", and "*dangerous*" in the natural way. Then we can write the statements like so:

   (a) *seeker* ∨ *smart*,
   (b) *seeker* → *dangerous*,

(c) *smart* → *evil*,

(d) ¬*dangerous* → ¬*evil*.

The contrapositive of the last statement is *evil* → *dangerous*; combined with the third statement, this gives *smart* → *dangerous*, and we also have *seeker* → *dangerous*. Since at least one of *seeker* and *smart* is true, we can conclude ⟨Fluffy is dangerous⟩.

31. Write $n = \underline{abcd}$, so that $9n = \underline{dcba}$. Since $n \geq 1000$, we know $9n \geq 9000$ and so $d = 9$. Since $9n < 10000$, we know $n \leq 1111$; since $a \neq b$, this means $a = 1$ and $b = 0$. Finally, because $\underline{dcba}$ is a multiple of 9, so is $d + c + b + a$, therefore $c = 8$ and $n = \boxed{1089}$. (Indeed, $9 \cdot 1089 = 9801$.)

Level 3

32. The available digits are 2, 3, 5, and 7. If we fix the last digit of a number, there are $3! = 6$ ways for the other digits to be arranged, so there are 6 numbers ending in each digit. Thus the sum of the units digits of all the numbers is $6 \cdot (2 + 3 + 5 + 7) = 102$. The same analysis applies to all the tens digits; their sum is also 102, and so the amount that they contribute to the sum of the numbers is $10 \cdot 102$. The same thing again applies to the hundreds and thousands digits, so the total sum is $102 \cdot (1000 + 100 + 10 + 1) = \boxed{113322}$.

33. **Solution 1:** As in problem 22, the sum of the digits of all the numbers from 0 to 999 is 13500. If n is an even number, $n + 1$ is an odd number whose digit sum is greater by 1. There are 500 even and 500 odd numbers in the range, so the sum of all the digits of the odd numbers is 500 more than that sum for the even numbers. Thus the sum for the even numbers is $\frac{13500 - 500}{2} = 6500$, and adding the contribution of 1 from 1000 gives $\boxed{6501}$.

Solution 2: Any even number from 000 to 999 can be formed by writing an integer from 00 to 99 and following it with one of 0, 2, 4, 6, and 8. In the list of numbers from 00 to 99, there are 200 digits in total, with each digit appearing 20 times, so the sum of all the digits in the list is $20 \cdot (0 + \cdots + 9) = 900$. We must write this list 5 times, for a total digit sum of 4500, and then add an additional 100 copies each of 0, 2, 4, 6, and 8, for a further $100 \cdot (0 + 2 + 4 + 6 + 8) = 2000$. Finally, we add the contribution from 1000 to get $4500 + 2000 + 1 = \boxed{6501}$.

34. We can get any multiple of 25 in the range, except for 10^5 itself, by writing an integer from 000 to 999 and following it with one of 00, 25, 50, and 75. As in problem 22, the sum of the digits of the numbers from 000 to 999 is 13500. As in the previous problem, we must multiply this by 4, giving 54000, then add $1000 \cdot \big((0+0) + (2+5) + (5+0) + (7+5)\big) = 24000$, and finally add 1 to get $\boxed{78001}$.

35. By direct calculation, you can see that the units digits of the powers of any integer form a cycle of length 1, 2, or 4, so the fifth power of any integer has the same units digit as the integer itself. (Readers familiar with Fermat's Little Theorem may also note that it implies $a^5 \equiv a$ for any a, modulo both 2 and 5, and therefore modulo 10.) In the integers from 1 to 2000, each digit appears as a units digit 200 times, so the sum of those integers has a units digit of 0. We just need the units digit of $2001^5 + 2002^5 + \cdots + 2006^5$ which is the units digit of $1 + 2 + 3 + 4 + 5 + 6 = 21$, and that digit is $\boxed{1}$.

Chapter 2

Algebra

Questions

Level 1

1. Find the largest integer x such that $\frac{24}{x+3}$ is an integer.
2. A seven-sided polygon has three sides of length $x^2 + y^2$ and four sides of length $x^2 - y^2$. Write an expression for the perimeter of the polygon in terms of x and y.
3. If the average of x and $4x$ is 20, compute x.
4. Grace got an average score of 80 on her first four tests. Her score on the 5th test was 16 points lower than the average of all five tests. Compute her score on the 5th test.
5. Find the smallest integer $n > 1$ such that the sum $1 + 2 + \cdots + n$ is a perfect square.
6. If $x - 432$ is 10 more than $y + 568$, compute $x - y$.
7. Eating at a constant rate, Sonny can eat 3 pies in 4 minutes. At this rate, in how many minutes can he eat 5 pies?
8. Two sides of a square have lengths x and $4x - 12$. Find the numerical value of the area of the square.
9. A toy boat costs $\frac{3}{4}$ of its price plus 15 cents. Compute the price of the boat.
10. One third of a number plus twice half of the number equals 36. Compute the number.
11. A tank of oil is $\frac{1}{6}$ full. After 75 gallons of oil are removed, it is $\frac{1}{7}$ full. How many gallons of oil does the tank hold when it is full?

12. When a quantity of water freezes, it increases in volume by $\frac{1}{11}$. When a quantity of ice melts, by what fraction does its volume decrease?

13. Ten percent of 9 is 9 percent of what number?

14. Annie is collecting wild berries. She finds that the berries lose 10% of their weight as they dry out on the way home. Compute how many pounds she must collect to arrive home with 100 pounds of berries.

15. A dealer buys a car from the factory for 20% less than the sticker price. He sells the same car for the full sticker price. What percent of his cost is his profit?

16. A white dress sold originally for $50, and was marked down 10% for a sale. After being marked up 10%, a blue dress sold for the same price as the sale price of the white dress. To the nearest dollar, compute the original price of the blue dress.

17. A discount clothing store bought a pair of pants for 34% less than the price on the tag. The store sold the pants for 25% less than the price on the tag. What percent of the final price was profit for the store?

18. One quarter is the same part of one third as one half is of what number?

19. Compute the numerical value of
$$2^{\left(2^{\left(2^2\right)}\right)} - \left(\left(2^2\right)^2\right)^2.$$

20. If $x = -2$, compute $-(-x - x^x)^{-x}$.

21. If $\frac{q \cdot q \cdot q}{q + q + q} = 48$ and $q > 0$, compute q.

22. Compute the numerical value of $\sqrt{5^5 + 5^5 + 5^5 + 5^5 + 5^5}$.

23. If 2^y is four times 8^x, write an equation expressing y in terms of x.

24. If $10^x = 5$, compute 10^{2x+1}.

25. If $4^x = \sqrt{2^{3y}}$, compute $\frac{x}{y}$.

26. If $7^{3x} = 216$, compute 7^{-x}.

27. Compute all real values of x for which $2^{x^2} = (2^x)^2$.

28. One third of a certain positive number equals five times the reciprocal of the number. Find the number.

29. Find all real numbers x such that $x - 1$ is the reciprocal of $x + 1$.

30. Without doing any squaring, express $\sqrt{313^2 - 312^2}$ as a positive integer.

31. Find the length of the edge of a cube if its volume in cubic units is numerically equal to its surface area in square units.

32. In the Bronx Zoo, the ratio of snakes to lizards is 3:2 and the ratio of lizards to monkeys is 5:3. Compute the ratio of snakes to monkeys.

33. If the sum of two numbers is 20 and their product is 30, compute the sum of their reciprocals.

34. Several people are walking their dogs. If the total number of legs among both dogs and humans is 22 more than twice the total number of heads, compute the number of dogs.

35. Larry leaves the house at noon and travels east at 40 miles per hour. At 2:30 p.m., his wife Joanne leaves, taking the same route and traveling at 60 miles per hour. At what time will Joanne overtake Larry?

36. Compute the remainder when $x^{10} + x^5 + 1$ is divided by $x + 1$.

37. Winnifred invests K dollars in a bond yielding $6\frac{1}{2}\%$ simple interest and $K + 1000$ dollars in a bond yielding 9% simple interest. If the interest from the second bond is \$102 less than the twice the interest from the first bond, find K.

38. Find all real numbers x such that $x^4 - 13x^2 + 36 = 0$.

39. If $a{:}b{:}c = 2{:}3{:}4$, compute $\frac{a}{b} \div \frac{b}{c}$.

40. A certain map has a scale in which one inch represents 20 miles. If the area of the county is 600 square miles, compute the area in square inches of the region representing this county on the map.

41. Compute all x such that $\frac{x-2}{x-1} = \frac{x+1}{x+3}$.

42. Sending a telegram costs m cents for the first 12 words and s cents for each additional word. If $k > 12$, express in terms of m, k, and s the cost of sending a telegram of k words.

43. Compute $\sqrt[3]{\sqrt{30} + \sqrt{3}} \cdot \sqrt[3]{\sqrt{30} - \sqrt{3}}$.

44. A right circular cone and right circular cylinder have equal heights and equal volumes. Compute the ratio of the base radius of the cylinder to the base radius of the cone.

45. A girl has only red, blue, and green stickers. She has $\frac{1}{3}$ as many blue stickers as red stickers, and $\frac{1}{6}$ as many red stickers as green stickers. Compute the fraction of her sticker collection that consists of green stickers.

46. How many ounces of pure water should be added to 20 ounces of a 45% acid solution to make it 30% acid?

47. A store makes a profit of 20% on an item, based on its selling price. What percent profit does the store make based on the cost?

48. The price of a pen was increased by $P\%$. The new price was decreased by $P\%$. If the final price was one dollar, express the original price of the pen in terms of P.

49. Find all roots of the equation $\sqrt{3-x} = x\sqrt{3-x}$.

50. Compute the real value of $x+y$ if $x^3 + 3x^2y + 3xy^2 + y^3 = -27$.

51. Find all real values of y satisfying $|\sqrt{y} - \sqrt{3}| < 1$.

52. A wooden cube with volume 64 cubic inches is painted on all six faces and then cut into 64, 1-inch cubes. How many of these cubes have exactly two faces painted?

53. A wooden cube with volume n^3 cubic inches is painted and then cut into n^3 identical cubes. Compute n if exactly 180 of the small cubes have exactly two faces painted.

54. Find the number of terms in $((x+3y)^2 \cdot (x-3y)^2)^2$ after it is simplified.

55. Compute the coefficient of the sixth term in the result when $(x-2)^6$ is expanded and arranged in decreasing powers of x.

56. Find the ordered pair of real numbers (a, b) such that $(2+i)^5 = a + bi$.

57. The infinite repeating decimal $2.\overline{72}$ is equal to $\frac{a}{b}$, where a and b are integers with no common factor other than 1. Compute $a + b$.

Level 2

58. If $a = \sqrt{3}$, $b = \sqrt[3]{5}$, and $c = \sqrt[4]{7}$, arrange a, b, and c in increasing order.

59. Compute the value of $\frac{1}{1991} + \frac{1992 \cdot 1990}{1991} - 1992$.

60. If $x = 1.21$, compute the value of

$$\frac{\sqrt{(x+2)^2 - 8x}}{\sqrt{x} - \frac{2}{\sqrt{x}}}.$$

61. If a and b are positive integers such that $a^2 + 24 = b^2$, compute the largest possible value of $a + b$.

62. What is the prime factorization of $13^3 + 7^3$?

63. In terms of p and q, find the sum of the reciprocals of the roots of the equation $x^2 + px + q = 0$.

64. If $i = \sqrt{-1}$, compute $i^1 + i^2 + i^3 + \cdots + i^{100}$.

65. Compute $\lfloor \frac{8000}{\sqrt{2}} \rfloor$.

66. Find all x such that $\lfloor \frac{3x}{2} \rfloor - 1 = 5$.

67. Find the smallest x such that $\lfloor x \rfloor + \lfloor 2x \rfloor + \lfloor 3x \rfloor + \lfloor 4x \rfloor = 15$.
68. Find the smallest x such that $\lfloor x \rfloor + \lfloor 3x \rfloor + \lfloor 5x \rfloor = 21$.
69. The numeral $\underline{6}\,\underline{a}\,\underline{4}$ is interpreted in base 9. If the number it represents is a multiple of 8, find the base-9 digit a.
70. If $8_{10} \cdot 8_{10} = 54_n$, where a_b means the interpretation of the digits of a as a base-b numeral, find the base-10 representation of 84_n.
71. If the first three terms of a geometric sequence are $\sqrt{2}$, $\sqrt[4]{2}$, and $\sqrt[12]{2}$, compute the fourth term.
72. The first term of an arithmetic sequence is r and the common difference is $2r$. Express the sum of the first n terms of the sequence in terms of r and n.
73. The average of n consecutive integers, of which the smallest is n, is 94. Find n.
74. Find the value of x if $\frac{1}{3}$, $\frac{1}{x}$, and $\frac{1}{5}$ form an arithmetic sequence.
75. The measures of the interior angles of a pentagon form an arithmetic sequence, with a common difference of $10°$. Compute the degree measure of the smallest interior angle of the pentagon.
76. Find the two roots of

$$\left(x - \frac{1}{4}\right)\left(x - \frac{1}{4}\right) + \left(x - \frac{1}{4}\right)\left(x - \frac{1}{8}\right) = 0.$$

77. Find the distance between the two points of intersection of the graphs of $x^2 + y = 7$ and $x + y = 7$.
78. If two roots of $x^3 + px + q = 0$ are -1 and 3, find the third root in terms of p and q.
79. If the points $(2, 3)$, $(-1, 9)$, and $(6, k)$ are collinear, compute k.
80. Compute all x such that $2^x + 2^{x-1} = 48$.
81. Compute all x such that $x^2 - (1 + \sqrt{2})x + \sqrt{2} = 0$.
82. Compute the ordered pair of real numbers (a, b) such that $\frac{1}{1+i} + \frac{2}{a+bi} = \frac{1}{1-i}$.
83. In a sequence of natural numbers, the first two terms are consecutive integers, and each term after the second is one more than the sum of all the previous terms. If the tenth number is 1280, find the first number.
84. If $a_1 = 400$ and a_n is the sum of the cubes of the digits of a_{n-1} for all $n > 1$, find a_{1985}.
85. Find the smallest integer value of x for which $x^2 - 30x - 175$ is equal to a positive prime number.

86. Compute x if

$$\frac{58}{15} = 3 + \frac{1}{1 + \frac{1}{6 + \frac{1}{x}}}.$$

87. If $\frac{a}{b} = 5$ and $\frac{b}{c} = 6$, compute $\frac{a+2b+3c}{2a+3b+4c}$ as a fraction in lowest terms.

88. For what positive integral base b does $21_b \cdot 54_b = 1354_b$?

89. The number 12! is computed and converted to base 12. How many zeros does the number end with?

90. Find all x such that $|x|^2 + |x| - 42 = 0$.

91. Find the roots of the equation $x \cdot \lfloor x \rfloor = 40$.

92. When the polynomial $x^2 + bx + a$ is divided by $x + a$, the quotient is $x + a$ and the remainder is -6. If a and b are real numbers, find all possible values of b.

93. Compute all values of $\frac{1}{x}$ such that $2 - \frac{7}{x} + \frac{6}{x^2} = 0$.

94. Find the smallest positive three-digit integer k such that $\frac{k^5+k+1}{k^2+k+1}$ is an integer.

95. Find all positive integers x such that $\frac{x^3-98}{x-5}$ is an integer.

96. Find the sum of all positive integers n such that $n < 10$ and $n! + (n+1)!$ is divisible by 5.

97. Find the first time after 6:00 that the hands of a clock are perpendicular.

98. Two candles of the same length are lit at the same time and burn down uniformly. One candle takes 4 hours to burn completely, and the other takes 5 hours. How many hours after the candles are lit will one candle be 5 times as long as the other?

99. Find all ordered pairs of real numbers (a, b) which satisfy $(a+b)^2 = (a+1)(b-1)$.

100. Let $f(x) = x^2 + 1990x + 36$. How many pairs of positive integers (a, b) are there such that $f(a + b) = f(a) + f(b)$?

101. A man can paint a room in 3 hours, and his son can paint the room in 4 hours. How long would it take them to paint the room working together?

102. A newspaper office has two presses. The old one can print the newspaper in 6 hours, and the new one can print the newspaper in 5 hours. After they have both been working for 2 hours, the old one breaks down. How much longer does it take the new one to finish the job?

103. If a and b are the roots of $5x^2 + 6x + 7 = 0$, compute $\frac{1}{a} + \frac{1}{b}$.

104. If a and b are the roots of $3x^2 + 4x + 5 = 0$, compute $\frac{a}{b} + \frac{b}{a}$.

105. If a and b are the roots of $2x^2 + 3x + 4 = 0$, compute $\frac{a^2}{b} + \frac{b^2}{a}$.

106. Find all C such that the sum of the cubes and the sum of the squares of the roots of $Cx^2 + 4x + 3 = 0$ are equal.

107. Find all ordered pairs (m, n) such that the roots of $x^2 + mx + n = 0$ are $\frac{m}{3}$ and $\frac{n}{5}$.

108. Find the polynomial with leading coefficient 1 whose zeroes are each of the zeroes of $x^3 + 2x - 2$ multiplied by 3.

109. Compute the ordered pair (a, b) such that the zeroes of $x^2 + ax + b$ are the squares of the zeroes of $x^2 - 6x + 10$.

110. Find all ordered pairs of integers (x, y) which satisfy $2^{2x} - 3^{2y} = 55$.

111. Find the ordered pair of integers (a, b) such that $z = a \pm \sqrt{b}$ in the solutions to the system of equations

$$x + \frac{1}{y} = 1, \quad y + \frac{1}{z} = 2, \quad z + \frac{1}{x} = -3.$$

112. Find all integers n such that $(x - n)(x - 17) + 4 = 0$ has only integral roots.

113. Suppose t is a root of $x^4 - x^3 + x^2 - x + 1 = 0$. Compute $t^{20} + t^{10} + t^5 + 1$.

114. Let P be the set of all polynomials with integer coefficients whose coefficients sum to 0. Find the smallest integer $t > 1$ such that t divides $f(1992)$ for all f in P.

115. George drives to a distant city and returns using the same route. He must average 40 miles per hour over the whole trip in order to keep an appointment in his home city. He is delayed en route, and can only average 30 miles per hour on the trip to the distant city. Find the rate he must average on the return trip to keep the appointment.

116. In a class, there are 32 students who speak Mandarin, Cantonese or Toishanese. No one speaks only Cantonese or only Toishanese. The number who speak only Mandarin is equal to the number speaking only Cantonese and Mandarin. The number speaking only Cantonese and Toishanese is 4 times the number speaking all three. (At least one person speaks all three.) There are 4 who speak only Toishanese and Mandarin. If the number speaking only Cantonese and Toishanese is less than the number speaking only Mandarin and Cantonese, how many students speak only Mandarin?

117. Compute the product of all x such that $\frac{4}{x} - \frac{5}{x^3} + \frac{1}{x^5} = 0$.

118. Dauntless runs twice as fast as he walks. One day, on the way to school, he walks for twice the time he runs, and it takes him 20 minutes to get there. A week later, keeping the same pace, he runs for twice the time he walks. How many minutes does it take Dauntless to get to school the second time?

119. Compute $\frac{1}{7} + \frac{2}{7^2} + \frac{1}{7^3} + \frac{2}{7^4} + \frac{1}{7^5} + \cdots$.

120. If $A = 1 + \frac{1}{a} + \frac{1}{a^2} + \cdots$ and $B = 1 + \frac{1}{b} + \frac{1}{b^2} + \cdots$, find the ordered pair of positive integers (a, b) such that $a > b > 1$ and $A + B = \frac{15}{7}$.

121. The sum of an infinite geometric sequence is 4 and the sum of the first two terms is 3. Find all possible values of the first term.

122. Suppose that $x + \frac{1}{x} = i$, where $i = \sqrt{-1}$. Compute $x^4 + \frac{1}{x^4}$.

123. If $x + \frac{1}{x} = 10$, compute $\left| x - \frac{1}{x} \right|$.

124. Suppose that $x + \frac{1}{x} = -1$ and let $f(p) = x^p + \frac{1}{x^p}$. Compute

$$\sum_{n=1}^{1990} f(2^n).$$

125. Find the ordered pair (a, b) which satisfies

$$\sum_{x=1}^{10} ax^2 + b = \sum_{x=11}^{20} bx + a = 11915.$$

126. Express $\frac{1}{\sqrt[3]{7}-1}$ as a simple fraction with rational denominator.

127. Find the coefficient of x^{499} in the expansion of the product

$$(x - 2)(x - 4)(x - 6) \cdots (x - 1000).$$

128. Find all x such that $2 \lfloor 2x \rfloor^2 - 9 \lfloor 2x \rfloor + 9 = 0$.

129. Find all x such that $\lfloor 2x \rfloor^2 = x + \frac{3}{2}$.

130. Working alone, Tony can paint a room in 10 hours. After he has been working for 4 hours, Maria arrives, and they complete the job together in 2 additional hours. How long would it take Maria to paint the room alone?

131. Dan is 5 times as old as his daughter Natalie, but 21 years from now he will only be twice as old as she will be then. How old is he now?

132. A bus company currently charges $1 per ride and has 500 riders. It finds that for every nickel it raises the fare, it loses 10 riders. How much should it charge for maximum revenue?

133. At a snack bar, the cost of 7 sandwiches, 5 drinks, and 1 dessert is $32, while the cost of 10 sandwiches, 7 drinks, and 1 dessert is $45. Find the cost of 1 sandwich, 1 drink, and 1 dessert.

134. A student mistakenly thought that $\sqrt{5\frac{2}{3}} = 5\sqrt{\frac{2}{3}}$. Find all integers that can replace the 5 in that equation so that it actually becomes true.

135. Find the smallest integer x such that $\frac{1}{2} - \sqrt{x - 19} + \sqrt{x - 89} > 0$.

136. Compute the greatest real number x such that

$$\sqrt{x - \sqrt{2x - 1}} + \sqrt{x + \sqrt{2x - 1}} = \sqrt{2}.$$

137. In a class of 35 students, 20 own calculators, 11 own slide rules, and 10 own neither. Find how many students own both calculators and slide rules.

138. A driver finishes a trip of 70 miles at a constant rate in 2.5 hours. By how many miles per hour must he increase his speed to make the same trip $\frac{3}{4}$ hours faster?

139. Find all real x such that $3^x + 3^{x+1} = 324$.

140. The sum of two numbers is 12 and the difference of their squares is 96. Compute the larger of the two numbers.

141. If an object is released from a point above the surface of the earth and falls freely, after t seconds it has fallen $16t^2$ feet. Two objects are dropped at the same time, the first from a height of 100 feet and the second from a height of 76 feet. In how many seconds will the first object be twice as high from the ground as the second?

142. Find the ordered pair of natural numbers (a, b) such that $a + b\sqrt{3} = \sqrt[3]{26 + 15\sqrt{3}}$.

143. Willow owns a magic purse that doubles any sum of money placed into it. Willow charges $1.20 for each use of the purse. Alex starts with a certain sum of money, puts it in the purse to double it, and, after paying Willow his fee, reinvests all his money in the purse. After doubling his money again, and paying Willow the second fee, Alex reinvests all his money once more. After the money doubles and the third fee is paid, Alex finds that he has no money left at all. How much did Alex start with?

144. A club found that it could achieve a membership ratio of 2 men for each woman either by either recruiting 24 men or expelling x women. Compute x.

145. Find all real numbers x for which $(x + 2)^4 + x^4 = 82$.

146. Find all values of x such that $|x| = |7 - 2x| + 3$.

147. If a, b, and c are positive, $ab = 21$, $bc = 35$, and $ac = 15$, compute $a + b + c$.

148. Two students begin walking from point A to point B at the same constant rate and at the same time. One student walks with steps that are 60 centimeters apart, while the other walks with steps that are 69 centimeters apart. Their footsteps coincide when they first begin, and the next time they coincide is 15 seconds later. After 5 minutes of walking, they arrive at point B. Compute the distance from A to B in meters.

149. Find the value of n such that $\dfrac{\binom{2n}{n}}{n+1} = \dfrac{\binom{2n+2}{n+1}}{1990}$.

Level 3

150. Find an ordered pair of integers (a, b) such that

$$2(\sqrt{5} - \sqrt{3} + \sqrt{2}) = (\sqrt{5} + \sqrt{3} + \sqrt{2})(\sqrt{a} - \sqrt{b}).$$

151. Find all real values of x such that $\sqrt{3x^2 + 6x + 7} + \sqrt{5x^2 + 10x + 14} = 4 - 2x - x^2$.

152. Solve the following system of equations for (a, b, c) if a, b, and c are positive.

$$ab + bc + ca + a^2 = 182,$$
$$ab + bc + ca + b^2 = 294,$$
$$ab + bc + ca + c^2 = 273.$$

153. Find four ordered pairs of integers (x, y) such that $x^3 = y^3 + 217$.

154. For any positive x and y, let s be the smallest of the numbers x, $y + \frac{1}{x}$, and $\frac{1}{y}$. Compute the largest value of s that can be obtained in this way.

155. Compute $\frac{2}{1} + \frac{4}{9} + \frac{10}{81} + \frac{28}{729} + \frac{82}{6561} + \cdots$.

156. Find T if the zeroes of $19x^3 + 92x^2 + 23x + T$ form a geometric sequence.

157. If $a = \sqrt{1986} + \sqrt{1988}$, $b = 2\sqrt[4]{1986 \cdot 1988}$, and $c = 2\sqrt{1987}$, arrange a, b, and c in increasing order.

158. Compute $\sqrt{100 \cdot 101 \cdot 102 \cdot 103 + 1}$.

159. If $x + \frac{1}{x} = a$, express $x^5 + \frac{1}{x^5}$ in terms of a.

160. If $x = \sqrt[3]{6 + \sqrt[3]{6 + \sqrt[3]{6 + \cdots}}}$, compute all real x.

161. One of the roots of

$$((1+x)^2)^{\frac{1}{7}} + 3((1-x)^2)^{\frac{1}{7}} = 4(1-x^2)^{\frac{1}{7}}$$

is of the form $\frac{k}{1094}$, where k is a positive integer. Find k.

162. Find all ordered pairs of positive real numbers (x, y) such that $xy = 16$ and there exists a real z satisfying $x^{x+y} = y^{9z}$ and $y^{x+y} = x^z$.

Answers

Level 1

1. The number $x + 3$ must be a divisor of 24; the largest possible divisor is 24 itself, so $x + 3 = 24$ and $x = \boxed{21}$.

2. The perimeter is $3(x^2 + y^2) + 4(x^2 - y^2) = \boxed{7x^2 - y^2}$.

3. We have $20 = \frac{x+4x}{2} = \frac{5x}{2}$, so $5x = 40$ and $x = \boxed{8}$.

4. **Solution 1:** Since the average of the first four scores is 80, their sum must be 320. If we let the 5th score be x, then the average of all five is $\frac{320+x}{5}$. We have $x = \frac{320+x}{5} - 16$, and $x = \boxed{60}$.

   Solution 2: The 5th score drags the average down by $\frac{1}{5}$ of the distance from it to the average of the other scores, so $\frac{4}{5}$ of the distance is 16. Thus, the distance is 20 and the score is $\boxed{60}$.

5. Computing the sum for the possible values of n in order gives the sums $3, 6, 10, 15, 21, 28, 36$; the first square is 36, which occurs when $n = \boxed{8}$.

6. We are told that $x - 432 = y + 568 + 10$. Moving all the variables to one side and all the numbers to the other side gives $x - y = 432 + 568 + 10 = \boxed{1010}$.

7. It takes Sonny $\frac{4}{3}$ minutes to eat one pie, so it takes him $5 \cdot \frac{4}{3} = \boxed{\frac{20}{3}}$ minutes to eat 5 pies.

8. Since all the sides of a square are congruent, we have $x = 4x - 12$, so $3x = 12$ and $x = 4$. Thus, the area of the square is $4^2 = \boxed{16}$.

9. We have that 15 cents is $\frac{1}{4}$ of the cost of the boat, so the whole cost is 15 cents $\cdot 4 = \boxed{60 \text{ cents}}$.

10. If x is the number, then $36 = \frac{x}{3} + 2 \cdot \frac{x}{2} = \frac{x}{3} + x = \frac{4}{3}x$, so $x = \frac{3}{4} \cdot 36 = \boxed{27}$.

11. Since $\frac{1}{6} - \frac{1}{7} = \frac{1}{42}$ of the capacity of the tank is 75 gallons, the whole capacity is $42 \cdot 75$ gallons $= \boxed{3150 \text{ gallons}}$.

12. If I is the volume of some ice and W is the volume of the water that results when it melts, we are told that $I = W \cdot (1 + \frac{1}{11}) = \frac{12}{11}W$, so $W = \frac{11}{12}I = I \cdot (1 - \frac{1}{12})$, so the ice has to lose $\boxed{\frac{1}{12}}$ of its volume.

13. Let the number be x; since "percent" simply means "times 0.01", we have $10 \cdot 0.01 \cdot 9 = 9 \cdot 0.01 \cdot x$, so $x = \boxed{10}$.

14. If Annie collects b pounds of berries, she will have $0.9b = \frac{9}{10}b$ pounds left when she gets home. Thus, we want $\frac{9}{10}b = 100$, so $b = \boxed{\frac{1000}{9}}$.

15. If the sticker price of the car is s, then the dealer pays $0.8s$ for it. He sells it again for s, so his profit is $s - 0.8s = 0.2s$ and the percentage is $\frac{0.2s}{0.8s} = \boxed{25\%}$.

16. The white dress sold for $0.9 \cdot \$50 = \$45$. If the original price of the blue dress was x, it sold for $1.1x = \$45$, so $x = \frac{\$45}{1.1} = \frac{\$450}{11} = \$40\frac{10}{11}$, which rounds to $\boxed{\$41}$.

17. If the original price was p, the store bought for $0.66p$ and sold for $0.75p$. Its profit is $0.75p - 0.66p = 0.09p$, so the percentage is $\frac{0.09}{0.75} \cdot 100\% = \boxed{12\%}$.

18. If the number is x, we have $\frac{\frac{1}{4}}{\frac{1}{3}} = \frac{\frac{1}{2}}{x}$, and we can double the numerator and denominator of the left side to get $x = \boxed{\frac{2}{3}}$.

19.

$$2^{(2^{(2^2)})} - ((2^2)^2)^2$$
$$= 2^{(2^4)} - (2^4)^2$$
$$= 2^{16} - 2^8$$
$$= 2^8(2^8 - 1)$$
$$= 256 \cdot 255 = \boxed{65280}.$$

20.

$$-(2 - (-2)^{-2})^2 = -\left(2 - \frac{1}{4}\right)^2 = -\left(\frac{7}{4}\right)^2 = \boxed{-\frac{49}{16}}.$$

21. We have $48 = \frac{q^3}{3q} = \frac{q^2}{3}$, so $q^2 = 144$. Since $q > 0$, we must have $q = \boxed{12}$.

22. This expression equals $\sqrt{5 \cdot 5^5} = \sqrt{5^6} = 5^3 = \boxed{125}$.

23. $2^y = 4 \cdot 8^x = 2^2 \cdot 2^{3x} = 2^{2+3x}$, so $y = \boxed{3x+2}$.

24. $10^{2x+1} = (10^x)^2 \cdot 10^1 = 5^2 \cdot 10^1 = \boxed{250}$.

25. We have $2^{2x} = 2^{3y/2}$, so $2x = \frac{3y}{2}$ and $\frac{x}{y} = \boxed{\frac{3}{4}}$.

26. We have $216 = 6^3 = (7^x)^3$, so $7^x = 6$ and $7^{-x} = \frac{1}{7^x} = \boxed{\frac{1}{6}}$.

27. The right side equals 2^{2x}, so we want $x^2 = 2x$, which is satisfied by $x = \boxed{0 \text{ and } 2}$.

28. If x is the number, we have $\frac{x}{3} = \frac{5}{x}$, so $x^2 = 15$ and $x = \boxed{\sqrt{15}}$.

29. We have $x - 1 = \frac{1}{x+1}$, so $1 = (x-1)(x+1) = x^2 - 1$. Thus $x^2 = 2$ and $x = \boxed{\pm\sqrt{2}}$.

30. $\sqrt{313^2 - 312^2} = \sqrt{(313+312)(313-312)} = \sqrt{625} = \boxed{25}$.

31. If the edge length is e, then we have $e^3 = 6e^2$, so $e = \boxed{6}$.

32. We have $\frac{s}{l} = \frac{3}{2}$ and $\frac{l}{m} = \frac{5}{3}$, where s, l, and m are the numbers of snakes, lizards, and monkeys. Multiplying these equations gives $\frac{s}{m} = \boxed{\frac{5}{2}}$.

33. If we let the numbers be x and y, we have $\frac{1}{x} + \frac{1}{y} = \frac{x+y}{xy} = \frac{20}{30} = \boxed{\frac{2}{3}}$.

34. If there are d dogs p people, then (assuming everyone has all their legs) we have $4d + 2p = 22 + 2(d+p) = 22 + 2d + 2p$, so $2d = 22$ and $d = \boxed{11}$.

35. When Joanne overtakes him, Larry has traveled for x hours, and she has traveled for $(x - \frac{5}{2})$ hours. Using rate times time = distance, and noting that when his wife overtakes him they have traveled the same distance, we have $40x = 60(x - \frac{5}{2})$, so $x = 7.5$ and the time she overtakes him is at $\boxed{7{:}30}$.

36. Let $f(x) = x^{10} + x^5 + 1$; by the polynomial remainder theorem, the remainder when $f(x)$ is divided by $x+1$ is $f(-1) = \boxed{1}$.

37. Since interest equals principal multiplied by rate, we have $0.09 \cdot (K + \$1000) = 2 \cdot 0.065K - \$102$. Thus $9K + \$9000 = 13K - \$10200$, so $\$19200 = 4K$ and $K = \boxed{\$4800}$.

38. We can factor the given polynomial to get $(x^2 - 9)(x^2 - 4) = 0$. Thus $x^2 = 9$ or $x^2 = 4$, so $x = \boxed{\pm 2 \text{ or } \pm 3}$.

39. Since $\frac{a}{b} = \frac{2}{3}$ and $\frac{b}{c} = \frac{3}{4}$, we have $\frac{a}{b} \div \frac{b}{c} = \frac{2}{3} \div \frac{3}{4} = \frac{2}{3} \cdot \frac{4}{3} = \boxed{\frac{8}{9}}$.

40. A map is geometrically similar to the area it represents. Since the linear dimensions are in the ratio $1\,\text{in}{:}20\,\text{mi}$, the areas are in the ratio $(1\,\text{in})^2{:}(20\,\text{mi})^2 = 1\,\text{in}^2{:}400\,\text{mi}^2$. So, if x is the area on the map, we have $x{:}600\,\text{mi}^2 = 1\,\text{in}^2{:}400\,\text{mi}^2$, so $x = \boxed{1.5\,\text{in}^2}$.

41. Cross-multiplying gives $x^2 - 1 = x^2 + x - 6$, so $-1 = x - 6$ and $x = \boxed{5}$.

42. The cost consists of m cents for 12 of the words and then s cents for each of the remaining $k - 12$ words, so the total cost is $m + s \cdot (k - 12) = \boxed{m + ks - 12s}$.

43. $\sqrt[3]{\sqrt{30} + \sqrt{3}} \cdot \sqrt[3]{\sqrt{30} - \sqrt{3}} = \sqrt[3]{\sqrt{30}^2 - \sqrt{3}^2} = \sqrt[3]{27} = \boxed{3}$.

44. If the common height is h, the base radius of the cone is r, and the base radius of the cylinder is s, we have $\pi r^2 h = \frac{\pi s^2 h}{3}$, so $r^2 = \frac{s^2}{3}$ and $\frac{s}{r} = \boxed{\sqrt{\frac{1}{3}}}$.

45. Let r, g, and b be the number of red, green, and blue stickers respectively. We have $r = 3b$ and $g = 6r$, so $g = 18b$. The total number of stickers is $r + g + b = 22b$, and the proportion of green stickers is $\frac{18b}{22b} = \boxed{\frac{9}{11}}$.

46. Adding more water does not change the amount of acid present, so we need $0.45 \cdot 20 = 0.3 \cdot (20 + x)$, so $30 = 20 + x$ and $x = \boxed{10}$.

47. Let p be the profit, c the cost, and s the selling price. We have $s = c + p$ and $p = 0.2s$, so $c = 0.8s$ and $\frac{p}{c} = \frac{1}{4} = \boxed{25\%}$.

48. Let the original price be x and let $p = \frac{P}{100}$. Then the price after the increase is $x \cdot (1 + p)$ and the price after the decrease is $x \cdot (1 + p) \cdot (1 - p) = x \cdot (1 - p^2)$. This equals 1, so $x = \frac{1}{1 - p^2} = \boxed{\frac{10000}{10000 - P^2}}$.

49. Rewrite the equation as $\sqrt{3 - x}(1 - x) = 0$, so either $\sqrt{3 - x} = 0$, which means $x = 3$, or $1 - x = 0$, which means $x = 1$. Both of those check out, so the solutions are $\boxed{1 \text{ and } 3}$.

50. We have $(x + y)^3 = -27$, so $x + y = \boxed{-3}$.

51. The given equation is equivalent to $\sqrt{y} - \sqrt{3} \in (-1, 1)$, or $\sqrt{y} \in (\sqrt{3} - 1, \sqrt{3} + 1)$. Since both bounds are positive, we can square everything to get $\boxed{y \in (4 - 2\sqrt{3}, 4 + 2\sqrt{3})}$.

52. In order to have exactly two faces painted, a small cube must be on an edge, but not be a corner cube (which would have three). If the edge length of the cube is n inches, each edge has n cubes, of which 2 are corner cubes. Since a cube has 12 edges, the number of cubes painted on two sides is $12(n - 2)$. We have $n = \sqrt[3]{64} = 4$ here, so there are $12 \cdot 2 = \boxed{24}$ such small cubes.

53. From the previous problem we have $12(n - 2) = 180$, so $n - 2 = 15$ and $n = \boxed{17}$.

54. The given expression equals $(x^2 - 9y^2)^4$; the binomial theorem tells us that $(a + b)^n$ has $n + 1$ terms (if a and b do not combine with each other), so there are $\boxed{5}$ terms.

55. By the binomial theorem, the sixth term of $(a + b)^6$, ordered by decreasing powers of a, is $\binom{6}{1}ab^5 = 6ab^5$. In this case, $a = x$ and $b = -2$, so the term is $6x(-2)^5 = -192x$, and the coefficient is $\boxed{-192}$.

56. **Solution 1:** The binomial theorem gives

$$(2 + i)^5 = 2^5 + 5 \cdot 2^4 \cdot i + 10 \cdot 2^3 \cdot i^2 + 10 \cdot 2^2 \cdot i^3 + 5 \cdot 2 \cdot i^4 + i^5$$
$$= 32 + 80i - 80 - 40i + 10 + i$$
$$= \boxed{-38 + 41i}.$$

Solution 2:

$$(2 + i)^2 = 2^2 + 2 \cdot 2 \cdot i + i^2$$
$$= 3 + 4i$$
$$(2 + i)^4 = (3 + 4i)^2 = 9 + 2 \cdot 3 \cdot 4i + 16i^2$$
$$= -7 + 24i$$
$$(2 + i)^5 = (-7 + 24i)(2 + i)$$
$$= -14 - 7i + 48i + 24i^2$$
$$= \boxed{-38 + 41i}.$$

57. **Solution 1:** Let $x = 2.\overline{72}$. Then $100x = 272.\overline{72} = 270 + x$, so $x = \frac{270}{99} = \frac{30}{11}$, and the sum is $30 + 11 = \boxed{41}$.

Solution 2: We have $0.\overline{72} = 8 \cdot 0.\overline{09} = \frac{8}{11}$, so $x = 2 + \frac{8}{11} = \frac{30}{11}$, and the sum is $\boxed{41}$.

Level 2

58. We have
$$a^{12} = 3^6 = 729,$$
$$b^{12} = 5^4 = 625,$$
$$c^{12} = 7^3 = 343,$$
so $c^{12} < b^{12} < a^{12}$, and, since they are all positive, in increasing order we have $\boxed{c, b, a}$.

59. Let $n = 1991$, so we have
$$\frac{1}{a} + \frac{(a+1)(a-1)}{a} - (a+1) = \frac{1 + (a^2 - 1)}{a} - (a+1)$$
$$= a - (a+1) = \boxed{-1}.$$

60. Multiply numerator and denominator by $\sqrt{x}$ and expand inside the radical to get
$$\sqrt{x} \cdot \frac{\sqrt{x^2 + 4x + 4 - 8x}}{x - 2} = \sqrt{x} \cdot \frac{\sqrt{(x-2)^2}}{x - 2} = \sqrt{x} \cdot \frac{|x - 2|}{x - 2}.$$
We have $x - 2 < 0$, so $\frac{|x-2|}{x-2} = -1$, and also $\sqrt{x} = 1.1$. Thus the final result is $\boxed{-1.1}$.

61. We have $24 = b^2 - a^2 = (b-a)(b+a)$, so $b-a$ and $b+a$ are divisors of 24. Since b is their average, they must have the same parity, so 1 and 24 do not work, but 2 and $\boxed{12}$ do.

62. Since $x^3 + y^3 = (x+y)(x^2 - xy + y^2)$ in general, we have $13^3 + 7^3 = (13 + 7)(169 - 91 + 49) = 20 \cdot 127 = \boxed{2^2 \cdot 5 \cdot 127}$.

63. If the roots are r_1 and r_2, Vieta's formulas give $r_1 + r_2 = -p$ and $r_1 r_2 = q$; then we have $\frac{1}{r_1} + \frac{1}{r_2} = \frac{r_1 + r_2}{r_2} = \boxed{-\frac{p}{q}}$.

64. The powers of i occur in cycles of length 4: $i, -1, -i, 1$. These add to 0 and there are exactly 25 full cycles in the sum, so the sum evaluates to $\boxed{0}$.

65. We have
$$\left\lfloor \frac{8000}{\sqrt{2}} \right\rfloor = \left\lfloor \frac{8000\sqrt{2}}{2} \right\rfloor = \boxed{4000\sqrt{2}}.$$

Since $\sqrt{2} = 1.41421\ldots$, we have $1000\sqrt{2} = 1414.21\ldots$ and $5656 < 4000\sqrt{2} < 5657$, so $\lfloor 4000\sqrt{2} \rfloor = \boxed{5656}$.

66. We have $\lfloor \frac{3x}{2} \rfloor = 6$ and so $\frac{3x}{2} \in [6,7)$; multiplying by $\frac{2}{3}$ gives $\boxed{x \in [4, \frac{14}{3})}$.

67. Let $f(x) = \lfloor x \rfloor + \lfloor 2x \rfloor + \lfloor 3x \rfloor + \lfloor 4x \rfloor$. Every term in the sum, and therefore $f(x)$ as a whole, is a nondecreasing function of x. We have $f(2) = 20$, so $x < 2$. Also, $f(x) \le 10x$, so $x \ge \frac{15}{10} = \frac{3}{2}$, and $f(\frac{3}{2}) = 1 + 3 + 4 + 6 = 14$. As x changes, $f(x)$ can only change at a multiple of $1, \frac{1}{2}, \frac{1}{3},$ or $\frac{1}{4}$. The smallest such number greater than $\frac{3}{2}$ is $\frac{5}{3}$, and indeed $f(\frac{5}{3}) = 15$, so the smallest number is $\boxed{\frac{5}{3}}$.

68. Let $f(x) = \lfloor x \rfloor + \lfloor 3x \rfloor + \lfloor 5x \rfloor$. As in the previous problem, we have $f(3) = 27$, so $x < 3$, and $f(x) \le 9x$, so $x \ge \frac{21}{9} = \frac{7}{3}$, and $f(\frac{7}{3}) = 2 + 7 + 11 = 20$. $f(x)$ can only change when x is a multiple of $1, \frac{1}{3},$ or $\frac{1}{5}$. The next such number after $\frac{7}{3}$ is $\frac{12}{5}$, and indeed $f(\frac{12}{5}) = 21$, so the answer is $\boxed{\frac{12}{5}}$.

69. The divisibility test for 8 in base 9 works the same way as the divisibility test for 9 in base 10; that is, we want the sum of the digits to be a multiple of 8. The sum is $10 + a$, so we need $a = \boxed{6}$.

70. **Solution 1:** We have $8 \cdot 8 = 5x + 4$, so $x = 12$, and $84_x = 8 \cdot 12 + 4 = \boxed{100}$.

   Solution 2: We have $54_x = 64$, so $50_x = 60$ and $30_x = \frac{3}{5} \cdot 60 = 36$. Thus $84_x = 30_x + 54_x = 36 + 64 = \boxed{100}$.

71. The given terms are equal to $2^{1/4}, 2^{1/6},$ and $2^{1/12}$, or, equivalently, $2^{3/12}, 2^{2/12},$ and $2^{1/12}$. The common ratio is therefore $2^{-1/12}$, so the next term is $2^0 = \boxed{1}$.

72. The last term in the sequence is $r + 2r \cdot (n-1)$, so the sum of all n terms is $n \cdot \frac{r + r + 2r \cdot (n-1)}{2} = \boxed{rn^2}$.

73. The last term in the sequence is $2n - 1$, so the average of all the terms is $\frac{n + 2n - 1}{2} = \frac{3n-1}{2}$. Thus $\frac{3n-1}{2} = 94$ and $n = \boxed{63}$.

74. In an arithmetic sequence, each term is the arithmetic mean of its neighbors, so we have $\frac{1}{x} = \frac{\frac{1}{3} + \frac{1}{5}}{2} = \frac{\frac{8}{15}}{2} = \frac{4}{15}$, and so $x = \boxed{\frac{15}{4}}$.

75. The sum of the measures of all the angles is $180° \cdot (5 - 2) = 540°$. Let the measures of the angles be $a - 20°$, $a - 10°$, a, $a + 10°$, and $a + 20°$; their sum is $5a$, so $a = 108°$ and the smallest angle measures $\boxed{88°}$.

76. Factor the equation to get $\left(x - \frac{1}{4}\right)\left(x - \frac{1}{4} + x - \frac{1}{8}\right) = \left(x - \frac{1}{4}\right)$ $\left(2x - \frac{3}{8}\right) = 0$, so $x = \boxed{\frac{1}{4} \text{ or } \frac{3}{16}}$.

77. Subtracting the second equation from the first gives $x^2 - x = 0$, so $x = 0$ or $x = 1$. We have $y = 7 - x$, so the intersections are $(0, 7)$ and $(1, 6)$, and the distance between them is $\sqrt{1^2 + 1^2} = \boxed{\sqrt{2}}$.

78. Since the coefficient of x^2 is 0, the sum of the roots is 0, so the third root is $\boxed{-2}$.

79. The slopes of the segments connecting any two of the points must be the same; thus we have $\frac{k-9}{6-(-1)} = \frac{9-3}{-1-2}$, so $\frac{k-9}{7} = -2$, and $k = \boxed{-5}$.

80. We have $48 = 2^{x-1}(2 + 1) = 3 \cdot 2^{x-1}$, so $2^{x-1} = 16$ and $x = \boxed{5}$.

81. The sum of the two roots is $1 + \sqrt{2}$ and their product is $\sqrt{2}$; by inspection, the two roots are $\boxed{1 \text{ and } \sqrt{2}}$. (You can also use the quadratic formula, which gives the same answer after more computation.)

82. Subtract $\frac{1}{1+i}$ from both sides to get $\frac{2}{a+bi} = \frac{1}{1-i} - \frac{1}{1+i} = \frac{2i}{1-i^2} = \frac{2i}{2} = i$. Thus we have $2 = i \cdot (a + bi) = -b + ai$, so $(a, b) = \boxed{(0, -2)}$.

83. If the first term is a, then the sequence begins $a, a+1, 2a+2, 4a+4$. By induction, the nth term is $2^{n-2}(a + 1)$, for $n > 1$. Thus $1280 = 2^8(a + 1) = 256(a + 1)$, so $a = \boxed{4}$.

84. By direct computation, we have $a_2 = 64$, $a_3 = 280$, $a_4 = 520$, $a_5 = 133$, $a_6 = 55$, $a_7 = 250$, and $a_8 = 133$. Since $a_5 = a_8$, the values after that will repeat in a cycle of length 3. Since $1985 \equiv 8 \pmod 3$, we have $a_{1985} = a_8 = \boxed{133}$.

85. The given expression equals $(x - 35)(x + 5)$; for this to be prime, one of the factors must be 1 or -1, and the other must be a prime or negative of a prime, accordingly. The smallest possible value of x is thus the one that makes $x + 5 = -1$; then $x - 35 = -41$, a negative prime, so $x = \boxed{-6}$.

86. We have

$$\frac{58}{15} = 3 + \frac{13}{15} = 3 + \frac{1}{1 + \frac{2}{13}} = 3 + \frac{1}{1 + \frac{1}{\frac{13}{2}}} = 3 + \frac{1}{1 + \frac{1}{6 + \frac{1}{2}}},$$

so $x = \boxed{2}$.

87. We can assume that, for example, $c = 1$; then $b = 6c = 6$ and $a = 5b = 30$, so $\frac{a+2b+3c}{2a+3b+4c} = \frac{30+12+3}{60+18+4} = \boxed{\frac{45}{82}}$.

88. **Solution 1:** Expanding out the values of all the numbers in base b gives

$$b^3 + 3b^2 + 5b + 4 = (2b + 1)(5b + 4)$$
$$= 10b^2 + 13b + 4$$
$$0 = b^3 - 7b^2 - 8b$$
$$= b(b - 8)(b + 1).$$

The only positive solution is $b = \boxed{8}$.

Solution 2: Write out the long multiplication, ignoring potential carries:

$$
\begin{array}{r}
5\,4 \\
\times\ 2\,1 \\
\hline
5\,4 \\
+\ 10\,8 \\
\hline
1\ \ 3\,5\,4
\end{array}
$$

Looking at the second-rightmost column of the sum shows that $5 + 8 \equiv 5 \pmod{b}$, so b is a divisor of 8; we know $b > 5$, since 5 is a valid digit in base b, so $b = \boxed{8}$.

89. A zero will be produced by every factor of $12 = 2^2 \cdot 3$, so we need to compute how many powers of 2 are in the product. There are $\lfloor \frac{12}{2} \rfloor = 6$ multiples of 2, each contributing at least one power; of these, $\lfloor \frac{6}{2} \rfloor = 3$ are multiples of 4, contributing an additional power; of these $\lfloor \frac{3}{2} \rfloor = 1$ are multiples of 8, contributing an additional power. Since $\lfloor \frac{1}{2} \rfloor = 0$, there are no more. Thus there are $6+3+1 = 10$ powers of 2. Similarly, there are $\lfloor \frac{12}{3} \rfloor + \lfloor \frac{12}{3^2} \rfloor = 5$ powers of 3. Thus there are $\boxed{5}$ zeroes.

90. We can factor the left side to see $(|x| + 7)(|x| - 6) = 0$, so $|x| = -7$ or $|x| = 6$. But $|x| > 0$, by the definition of the absolute value, so $|x| = 6$ and $x = \boxed{\pm 6}$.

91. For positive x, both x and $\lfloor x \rfloor$ are positive and nondecreasing, so their product is as well. If $x = 6$, then $x \cdot \lfloor x \rfloor = 36$, and $x = 7$ gives 49. Thus we want x between 6 and 7, so $\lfloor x \rfloor = 6$ and $x = \frac{40}{6} = \boxed{\frac{20}{3}}$. (There are no negative solutions, since we would similarly need $\lfloor x \rfloor = -7$ and thus $x = -\frac{40}{7}$, but $\lfloor -\frac{40}{7} \rfloor = -6$.)

92. We have $x^2 + bx + a = (x + a)(x + a) - 6 = x^2 + 2ax + (a^2 - 6)$. The coefficients of corresponding terms must match, so $b = 2a$ and $a = a^2 - 6$. The second equation has roots $a = 3$ and $a = -2$, and the first equation gives $b = \boxed{6 \text{ or } -4}$.

93. Factoring gives $(2 - \frac{1}{x})(1 - \frac{2}{x}) = 0$, so $\frac{1}{x} = \boxed{\frac{2}{3} \text{ or } \frac{1}{2}}$.

94. Since $\frac{k^5 + k + 1}{k^2 + k + 1} = k^3 - k^2 + 1$, (by long division), it is always an integer when k is an integer. So we simply need the smallest three-digit integer, which is $\boxed{100}$.

95. We have $\frac{x^3 - 98}{x - 5} = x^2 + 5x + 25 + \frac{27}{x - 5}$. Thus we need $x - 5$ to divide 27, so $x - 5 = \pm 1, \pm 3 \pm 9$, or ± 27, and so $x = \boxed{2, 4, 6, 8, 14 \text{ or } 32}$.

96. We have $n! + (n + 1)! = n! + n!(n + 1) = n!(n + 2)$. This is divisible by 5 when $n!$ is, which is true when $n \geq 5$, or when $n + 2$ is, which occurs when $n = 3$ or $n = 8$. Thus the sum is $3 + 5 + 6 + 7 + 8 + 9 = \boxed{38}$.

97. At 6:00, the hands form an angle of $180°$. The minute hand takes 60 minutes to make a full revolution of $360°$, so it moves by $\frac{360°}{60} = 6°$ every minute. The hour hand takes 12 hours to make a full revolution, so it moves by $\frac{1}{2}^{\circ}$ every minute. Thus the angle between the hands decreases by $\frac{11}{2}^{\circ}$ every minute. We need it to decrease by $90°$, which takes $\frac{90^{\circ}}{\frac{11^{\circ}}{2}} = \frac{180}{11} = 16\frac{4}{11}$ minutes, so the time is $\boxed{6:16\frac{4}{11}}$.

98. Let the original length of the candles be 1. In one hour, the first candle decreases in length by $\frac{1}{4}$ and the second by $\frac{1}{5}$; in t hours, the first decreases by $\frac{t}{4}$ and the second by $\frac{t}{5}$. Thus $1 - \frac{t}{5} = 5 \cdot \left(1 - \frac{t}{4}\right) = 5 - \frac{5t}{4}$, so $4 = \left(\frac{5}{4} - \frac{1}{5}\right)t = \frac{21}{20}t$ and $t = \boxed{\frac{80}{21}}$.

99. Multiplying out and regrouping gives $a^2 + (b + 1)a + b^2 - b + 1 = 0$; treat this as a quadratic equation in a. Its discriminant is

$$(b + 1)^2 - 4(b^2 - b + 1) = b^2 + 2b + 1 - 4b^2 + 4b - 4 = -3(b - 1)^2.$$

For a to be real, this must be nonnegative, which can only happen when $b = 1$. Then the equation is $a^2 + 2a + 1 = 0$, so $a = -1$ and $(a, b) = \boxed{(-1, 1)}$.

100. Expanding out the definition of f gives

$$(a + b)^2 + \cancel{1990(a + b)} + \cancel{36} = a^2 + \cancel{1990a} + \cancel{36} + b^2 + \cancel{1990b} + 36$$
$$a^2 + 2ab + b^2 = a^2 + b^2 + 36$$
$$2ab = 36.$$

Thus we need $ab = 18$; since $18 = 2^2 \cdot 3^1$, it has $(2 + 1)(1 + 1) = 6$ factors, so there are $\boxed{6}$ ordered pairs.

101. **Solution 1:** In one hour, the man can complete $\frac{1}{3}$ of the job and his son can complete $\frac{1}{4}$ of it; together, they can complete $\frac{1}{3} + \frac{1}{4} = \frac{7}{12}$ of the job. Thus the whole job takes them $\boxed{\frac{12}{7} \text{ hours}}$.

Solution 2: Let x be the number of hours they work together. In one hour, the man completes $\frac{1}{3}$ of the job. In x hours, working alone, the man completes $\frac{x}{3}$ and the son $\frac{x}{4}$ of the job. Working together for x hours, they complete one job. So, $\frac{x}{3} + \frac{x}{4} = 1$. This leads to $4x + 3x = 12$, and $x = \boxed{\frac{12}{7}}$.

102. After two hours, the old machine will have done $\frac{2}{6}$ of the job, the new one, $\frac{2}{5}$ of the job. $\frac{2}{6} + \frac{2}{5} + \frac{x}{5} = 1$. So, $x = \boxed{\frac{4}{3}}$.

103. Vieta's formulas give $a + b = -\frac{6}{5}$ and $ab = \frac{7}{5}$, so $\frac{1}{a} + \frac{1}{b} = \frac{a+b}{ab} = \frac{-6/5}{7/5} = \boxed{-\frac{6}{7}}$.

104. Vieta's formulas give $a + b = -\frac{4}{3}$ and $ab = \frac{5}{3}$, so $\frac{a}{b} + \frac{b}{a} = \frac{a^2+b^2}{ab} = \frac{(a+b)^2 - 2ab}{ab} = \frac{(-4/3)^2}{5/3} - 2 = \frac{16}{15} - 2 = \boxed{-\frac{14}{15}}$.

105. Vieta's formulas give $a + b = -\frac{3}{2}$ and $ab = \frac{4}{2}$, so

$$\frac{a^2}{b} + \frac{b^2}{a} = \frac{a^3 + b^3}{ab} = \frac{(a+b)^3 - 3ab^2 - 3ba^2}{ab}$$
$$= \frac{(a+b)^3 - 3ab(a+b)}{ab}$$
$$= \frac{(-3/2)^3}{4/2} + 3 \cdot \frac{3}{2}$$

$$= -\frac{27}{16} + \frac{9}{2}$$

$$= \boxed{\frac{45}{16}}.$$

106. Let the roots be a and b; then Vieta's formulas give $a + b = -\frac{4}{C}$ and $ab = \frac{3}{C}$. Then

$$a^2 + b^2 = (a+b)^2 - 2ab = \frac{16}{C^2} - \frac{6}{C}$$
$$a^3 + b^3 = (a+b)^3 - 3ab(a+b) = -\frac{64}{C^3} + \frac{36}{C^2}.$$

Equating these and multiplying by C^3 gives

$$-64 + 36C = 16C - 6C^2$$
$$3C^2 + 10C - 32 = 0$$
$$(C - 2)(3C + 16) = 0.$$

Thus $C = \boxed{2 \text{ or } -\frac{16}{3}}$.

107. Vieta's formulas imply $\frac{m}{3} + \frac{n}{5} = -m$, so $n = -\frac{20}{3}m$, and $\frac{mn}{15} = n$, so $n = 0$ or $m = 15$. The two solutions are therefore $(m,n) = \boxed{(0,0) \text{ or } (15,-100)}$.

108. To multiply the zeroes of a cubic polynomial by 3, we must (a) multiply the x^2 term by 3, since the sum of the roots is multiplied by 3; (b) multiply the x term by 3^2, since its coefficient is a sum of products of two roots at a time; and (c) multiply the constant term by 3^3, since it is (the negative of) the product of all three roots. Thus $x^3 + 2x - 2$ becomes $x^3 + 9 \cdot 2x - 27 \cdot 2 = \boxed{x^3 + 18x - 54}$.

109. Let the roots of the known equation be r and s; then the roots of the unknown equation are r^2 and s^2. Vieta's formulas on the known equation give $r + s = 6$ and $rs = 10$; on the unknown equation, they then give $b = r^2s^2 = (rs)^2 = 100$ and $-a = r^2 + s^2 = (r+s)^2 - 2rs = 16$. Thus $(a,b) = \boxed{(-16, 100)}$.

110. Factoring gives $(2^x + 3^y)(2^x - 3^y) = 55$. The first factor is positive and greater than the second, and the only ways to factor 55 are

$11 \cdot 5$ and $55 \cdot 1$, so the possibilities are

$$2^x + 3^y = 11 \qquad\qquad 2^x + 3^y = 55$$
$$2^x - 3^y = 5 \qquad\qquad 2^x - 3^y = 1$$
$$2 \cdot 2^x = 16 \qquad\qquad 2 \cdot 2^x = 56$$
$$2 \cdot 3^y = 6 \qquad\qquad 2 \cdot 3^y = 54$$

The second possibility leads to a non-integral value of x, and the first leads to $(x, y) = \boxed{(3, 1)}$.

111. The second equation gives $y = 2 - \frac{1}{z} = \frac{2z-1}{z}$, and the first then gives $x = 1 - \frac{1}{y} = 1 - \frac{z}{2z-1} = \frac{z-1}{2z-1}$. Then the third equation gives

$$-3 = z + \frac{2z-1}{z-1}$$
$$-3z + 3 = (z^2 - z) + (2z - 1)$$
$$0 = z^2 + 4z - 4.$$

The quadratic formula then gives $z = \frac{-4 \pm \sqrt{32}}{2} = -2 \pm \sqrt{8}$, so $(a, b) = \boxed{(-2, 8)}$.

112. We want $(x - n)(x - 17) = -4$; since x, n, and 17 are integers, so are $x - n$ and $x - 17$, so the possibilities are

$x - n$	$x - 17$	x	n
-4	1	18	22
-2	2	19	21
-1	4	23	22
1	-4	13	12
2	-2	15	13
4	-1	16	12

and the values of n leading to integral roots are $\boxed{12, 13, 21, \text{ and } 22}$. (If $n = 13$ or $n = 21$, the one corresponding value of x is a double root.)

113. Since t is a root, we know $t^4 - t^3 + t^2 - t + 1 = 0$. Multiplying by $t - 1$ gives $t^5 + 1 = 0$, so $t^5 = -1$. Thus

$$t^{20} + t^{10} + t^5 + 1 = (t^5)^4 + (t^5)^2 + t^5 + 1 = (-1)^4 + (-1)^2 + -1 + 1 = \boxed{2}.$$

114. The sum of the coefficients of any polynomial f equals $f(1)$, so 1 is a zero of every element of P; that is, $x - 1$ is a factor of every

element of P. Substituting $x = 1992$ gives that 1991 is a factor of $f(1992)$ for every f in P. Thus any divisor of 1991 is also a factor, and the smallest divisor besides 1 is $\boxed{11}$.

115. Let T be the time needed to make a one-way trip at 40 miles per hour; George needs to make both trips in $2T$. The trip out takes $T \cdot \frac{40}{30} = \frac{4}{3}T$, so he has $\frac{2}{3}T$ remaining; thus he must average double the speed on the return, which is $\boxed{60}$ miles per hour.

116. We can draw the Venn diagram below (Fig. 2.1) based on the given information, where x represents the number of students who speak only Mandarin and y represents the number who speak all three. We are told $4y < x$. Since there are 32 students in total, we have $32 = x + x + y + 4y + 4$, or $2x + 5y = 28$. The only integral solutions with positive y are $(x, y) = (4, 4)$ or $(x, y) = (9, 2)$; since we need $4y < x$, we are left with only $x = \boxed{9}$.

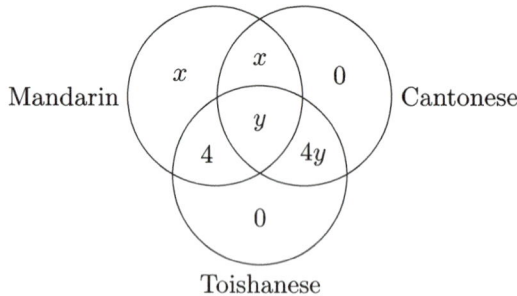

Figure 2.1

117. Factoring gives $\frac{1}{x}\left(4 - \frac{1}{x^2}\right)\left(1 - \frac{1}{x^2}\right) = 0$. The first factor cannot equal 0, and setting each of the other factors to 0 gives the solutions $x = \pm 1$, or $\pm\frac{1}{2}$ and the product is $\boxed{\frac{1}{4}}$.

118. Remember that distance is average rate multiplied by time. On the first day, Dauntless covers as much distance running as he does walking; thus the whole trip, relative to the walking portion, is 2 times the distance and $\frac{3}{2}$ times the time, so his average speed is $\frac{2}{3/2} = \frac{4}{3}$ of his walking speed. On the second day, the whole trip is 5 times the distance and 3 times the time, so his average speed is $\frac{5}{3}$ of his walking speed. Thus the second day takes $\frac{4/3}{5/3} = \frac{4}{5}$ as much time as the first day, which is $\frac{4}{5} \cdot 20 = \boxed{16}$ minutes.

119. We can rearrange the terms and use the formula for the sum of the terms of an infinite geometric sequence to get

$$\left(\frac{1}{7} + \frac{1}{7^2} + \frac{1}{7^3} + \cdots\right) + \left(\frac{1}{7^2} + \frac{1}{7^4} + \frac{1}{7^6} + \cdots\right)$$

$$= \frac{\frac{1}{7}}{1 - \frac{1}{7}} + \frac{\frac{1}{7^2}}{1 - \frac{1}{7^2}}$$

$$= \frac{1}{6} + \frac{1}{48}$$

$$= \boxed{\frac{3}{16}}.$$

120. We have $A = \frac{1}{1 - 1/a} = \frac{a}{a-1}$ and, similarly, $B = \frac{b}{b-1}$. Then

$$\frac{15}{7} = A + B = \frac{a \cdot (b-1) + b \cdot (a-1)}{(a-1)(b-1)}$$

$$= \frac{2ab - a - b}{ab - a - b + 1}$$

$$15ab - 15a - 15b + 15 = 14ab - 7a - 7b$$

$$ab - 8a - 8b + 15 = 0$$

$$ab - 8a - 8b + 64 = 49$$

$$(a-8)(b-8) = 49.$$

Since we want $a > b$, we must have $a - 8 = 49$ and $b - 8 = 1$, so $(a, b) = \boxed{(57, 9)}$.

121. If we let the sequence be $(a, ar, ar^2, \ldots)$, then we have $a + ar = 3$ and

$$(a + ar) + (ar^2 + ar^3) + (ar^4 + ar^5) + \cdots = \frac{a + ar}{1 - r^2} = 4,$$

so $1 - r^2 = \frac{3}{4}$. (Alternately, you could divide $a(1 + r) = 3$ by $\frac{a}{1-r} = 4$.) Thus $r = \pm\frac{1}{2}$. If $r = \frac{1}{2}$, then $\frac{a}{\frac{1}{2}} = 4$, so $a = 2$; if $r = -\frac{1}{2}$, then $\frac{a}{\frac{3}{2}} = 4$, so $a = 6$. Thus the possibilities are $\boxed{2 \text{ or } 6}$.

122. Square the given equation to get $x^2 + 2 + \frac{1}{x^2} = -1$, so $x^2 + \frac{1}{x^2} = -3$. Square again to get $x^4 + 2 + \frac{1}{x^4} = 9$, so $x^4 + \frac{1}{x^4} = \boxed{7}$.

123. Square the given equation to get $100 = x^2 + 2 + \frac{1}{x^2}$, so $96 = x^2 - 2 + \frac{1}{x^2} = \left(x - \frac{1}{x}\right)^2$. Since, in general, $\sqrt{a^2} = |a|$, we have $\sqrt{\left(x - \frac{1}{x}\right)^2} = |x - \frac{1}{x}| = \sqrt{96} = \boxed{4\sqrt{6}}$.

124. Square the given equation to get $x^2 + 2 + \frac{1}{x^2}$, so $x^2 + \frac{1}{x^2} = -1$. Square that to get $x^4 + 2 + \frac{1}{x^4} = 1$, so $x^4 + \frac{1}{x^4} = -1$. By induction we then have $f(2^n) = x^{2^n} + \frac{1}{x^{2^n}} = -1$ for all integer $n \geq 0$. Thus $\sum_{n=1}^{1990} f(2^n) = \sum_{n=1}^{1990} -1 = \boxed{-1990}$.

125. In general, we have
$$\sum_{i=1}^{n} 1 = n, \quad \sum_{i=1}^{n} i = \frac{n(n+1)}{2}, \quad \sum_{i=1}^{n} i^2 = \frac{n(n+1)(2n+1)}{6}.$$
Therefore,
$$11915 = \sum_{x=1}^{10} ax^2 + b = \left(a\sum_{x=1}^{10} x^2 \right) + \left(b\sum_{x=1}^{10} 1 \right)$$
$$= a \cdot \frac{10 \cdot 11 \cdot 21}{6} + b \cdot 10$$
$$= 385a + 10b$$
$$11915 = \sum_{x=11}^{20} bx + a = \left(\sum_{x=1}^{20} bx + a \right) - \left(\sum_{x=1}^{10} bx + a \right)$$
$$= b \cdot \left(\frac{21 \cdot 20}{2} - \frac{11 \cdot 10}{2} \right) + a \cdot 10$$
$$= 10a + 155b.$$

So we have, after dividing by 5, that $2383 = 77a + 2b = 2a + 31b$. From this we have $29 \cdot 2383 = 31 \cdot (77a+2b) - 2 \cdot (2a+31b) = 2383a$, so $a = 29$; the right equality gives $75a = 29b$, so $(a, b) = \boxed{(29, 75)}$.

126. In general, $a^3 - b^3 = (a-b)(a^2 + ab + b^2)$ and therefore $\frac{1}{a-b} = \frac{a^2+ab+b^2}{a^3-b^3}$; we can take $a = \sqrt[3]{7}$ and $b = 1$ to see $\frac{1}{\sqrt[3]{7}-1} = \boxed{\frac{\sqrt[3]{49}+\sqrt[3]{7}+1}{6}}$.

127. **Solution 1:** The equation $(x-2)(x-4)(x-6) \cdots (x-1000) = 0$ has 500 factors, with roots $2, 4, 6, \ldots, 1000$. The lead term of the expansion is x^{500} and the coefficient of the term with x^{499} is the negative of the sum of the roots, or $-(2 + 4 + 6 + \cdots + 1000) = \boxed{-250500}$.

Solution 2: You can think of computing the expansion of a product like this as follows: choose one of the terms from inside each parenthesized term and multiply all the choices together. Do that for all possible choices of terms, and add all of the products together. We can get a product that contains x^{499} by choosing x in 499 of the terms and the constant in the other term. The total coefficient of x^{499} is therefore the sum of all the constant terms, which is

$-2 - 4 - \cdots - 1000 = -2 \cdot (1 + 2 + \cdots + 500) = -500 \cdot 501 = \boxed{-250500}$.

128. Factoring the given equation yields $(2 \lfloor 2x \rfloor - 3)(\lfloor 2x \rfloor - 3) = 0$. We cannot have $\lfloor 2x \rfloor = \frac{3}{2}$, since it must be an integer by definition. Thus $\lfloor 2x \rfloor = 3$, which means $2x \in [3, 4)$, so $\boxed{x \in [\frac{3}{2}, 2)}$.

129. The left side of the equation is an integer; for the right side to be one, we must have $x = \frac{m}{2}$ for some odd integer m. Then we have $\lfloor 2x \rfloor = \lfloor m \rfloor = m$, so the equation becomes $m^2 = \frac{m+3}{2}$, or $2m^2 - m - 3 = 0$. This has roots $m = \frac{3}{2}$ and $m = -1$, but m must be an odd integer, so we have $m = -1$ and $x = \boxed{-\frac{1}{2}}$.

130. Tony works for 6 hours in all, and thus completes $\frac{6}{10} = \frac{3}{5}$ of the job; Maria completes the other $\frac{2}{5}$ in 2 hours of work. Thus she can do $\frac{1}{5}$ of the job in 1 hour, so she could do the whole job in $\boxed{5}$ hours.

131. **Solution 1:** Let x be Natalie's age, so that Dan's current age is $5x$. In 21 years, their ages will be $x + 21$ and $5x + 21$; we are told $5x + 21 = 2(x + 21)$, which is satisfied by $x = 7$. Thus, Dan's current age is $5 \cdot 7 = \boxed{35}$.

    Solution 2: The difference in their ages remains constant over time. It is currently 4 times Natalie's age; in 21 years, it will equal Natalie's age. So, if x is Natalie's current age, we have $4x = x + 21$, so $x = 7$, and Dan's age is $5 \cdot 7 = \boxed{35}$.

132. **Solution 1:** Let n be the number of nickels by which the fare has been raised. Then the number of riders is $500 - 10n$ and the fare is $20 + n$ nickels. The revenue is the product of these, which is $(500 - 10n)(20 + n) = 10000 + 300n - 10n^2$. The graph of this function is a downward-facing parabola; its turning point, which gives its maximum value, occurs when $n = \frac{-300}{2 \cdot -10} = 15$. Thus the fare should be $\$1 + 15 \cdot \$0.05 = \boxed{\$1.75}$.

    Solution 2: As above, we have $500 - 10n$ riders when the fare is $20 + n$ nickels. Raising the fare causes the loss of 10 riders and thus $200 + 10n$ nickels in revenue, but an additional nickel from each of the $490 - 10n$ remaining riders, for a total change of $290 - 20n$ nickels. This is positive until $n = 15$, so the fare should be $\$1 + 15 \cdot \$0.05 = \boxed{\$1.75}$.

133. Subtracting the first given purchase from the second shows that the cost of 3 sandwiches and 2 drinks is $13; subtracting this twice from the first purchase shows that the cost of 1 of each is $\boxed{\$6}$.

134. We want

$$\sqrt{x + \frac{2}{3}} = x \cdot \sqrt{\frac{2}{3}}$$

$$x + \frac{2}{3} = \frac{2}{3}x^2$$

$$0 = 2x^2 - 3x - 2$$

$$= (2x + 1)(x - 2).$$

Since we only want integer solutions, $x = \boxed{2}$.

135. Rewrite the equation as $\frac{1}{2} + \sqrt{x - 89} > \sqrt{x - 19}$. Since both sides are non-negative, we can square them to get $\frac{1}{4} + x - 89 + \sqrt{x - 89} > x - 19$, or $\sqrt{x - 89} > 70 - \frac{1}{4}$. Squaring again gives $x - 89 > 4900 - 35 + \frac{1}{16}$, or $x > 4954 + \frac{1}{16}$. Since x is an integer, this means $x \geq \boxed{4955}$.

136. Squaring and dividing by 2 gives

$$1 = x + \sqrt{x^2 - (2x - 1)} = x + \sqrt{(x - 1)^2} = x + |x - 1|.$$

If $x \geq 1$, then $|x - 1| = x - 1$ and we want $1 = x + (x - 1)$, which is satisfied by $x = 1$; thus the largest solution is $x = \boxed{1}$. (Note that if $x \leq 1$, we have $1 = x + (1 - x) = 1$, which is satisfied by *all* possible values of x. There will be imaginary numbers involved in the calculation if $x < \frac{1}{2}$, but the final result is still correct.)

137. Let C be the set of students who own calculators and S be the set of students who own slide rules. We have $|C \cup S| = 35 - 10 = 25$, and also $|C \cup S| = |C| + |S| - |C \cap S| = 20 + 11 - |C \cap S|$. Thus $|C \cap S| = \boxed{6}$.

138. **Solution 1:** The driver's current rate is $\frac{70\,\text{mi}}{2.5\,\text{h}} = 28\,\text{mph}$, and his new rate must be $\frac{2.5}{2.5 - \frac{3}{4}} = \frac{10}{7}$ times his current rate, for an increase of $\frac{3}{7} \cdot 28\,\text{mph} = \boxed{12}\,\text{mph}$.

Solution 2: As in Solution 1, the driver's rate is 28 miles per hour. If his rate increases by x miles per hour, he travels for 1.75 hours and covers a distance of 70 miles. So $(28 + x)(1.75) = 70$, and $x = \boxed{12}$.

139. $324 = 3^x + 3 \cdot 3^x = 4 \cdot 3^x$, so $3^x = 81$ and $x = \boxed{4}$.

140. **Solution 1:** We have $a + b = 12$ and $a^2 - b^2 = (a-b)(a+b) = 96$.
Thus $a - b = 8$, so $(a, b) = \left(\frac{12+8}{2}, \frac{12-8}{2}\right) = \boxed{(10, 2)}$.

Solution 2: We have $b = 12 - a$ and then $96 = a^2 - b^2 = a^2 - (12-a)^2 = -144 + 24a$, so $a = 10$, and thus $(a, b) = \boxed{(10, 2)}$.

141. **Solution 1:**

$$100 - 16t^2 = 2(76 - 16t^2)$$
$$100 - 16t^2 = 152 - 32t^2$$
$$16t^2 = 52$$
$$t^2 = \frac{13}{4}$$
$$t = \boxed{\frac{\sqrt{13}}{2}}.$$

Solution 2: The first object will always be $100 - 76 = 24$ feet higher than the second, so we need to find when the second object is 24 feet above the ground, i.e., it has fallen $76 - 24 = 52$ feet.

Thus we want $16t^2 = 52$, so $t^2 = \frac{13}{4}$ and $t = \boxed{\frac{\sqrt{13}}{2}}$.

142. Cubing both sides gives

$$26 + 15\sqrt{3} = a^3 + 3a^2 b\sqrt{3} + 9ab^2 + 3b^3\sqrt{3}.$$

Since a and b are natural numbers, we can equate the radical and non-radical terms to get $3b^3 + 3a^2 b = 15$ and $a^3 + 9ab^2 = 26$. If $b \geq 2$, then $9ab^2 \geq 36$, so the second equation cannot be satisfied. Thus $b = 1$, and substituting into the first equation gives $a = 2$, and $(a, b) = \boxed{(2, 1)}$.

143. **Solution 1:** If Alex starts with x dollars, after doubling and paying Willow once, he has $2x - 1.2$ dollars. After doubling and paying again, he has $4x - 3.6$ dollars. After the third time, he has $8x - 8.4 = 0$ dollars. Thus he starts with $\frac{\$8.4}{8} = \boxed{\$1.05}$.

Solution 2: Imagine watching the process in reverse: three times, Willow pays Alex $1.20 and then takes away half of his money. He starts with nothing, so the amounts of money he has after each step are $1.20, $.60, $1.80, $.90, $2.10, $\boxed{\$1.05}$.

144. Suppose there are currently m men and w women. Then $m + 24 = 2w$, so $m = 2w - 24$, and $m = 2(w - x) = 2w - 2x$. Thus $2x = 24$ and $x = \boxed{12}$.

145. Let $y = x + 1$, so our equation becomes

$$82 = (y + 1)^4 + (y - 1)^4$$
$$= 2y^4 + 12y^2 + 2$$
$$0 = y^4 + 6y^2 - 40$$
$$= (y^2 + 10)(y^2 - 4).$$

Since y is real, we must have $y^2 = 4$, so $y = \pm 2$ and $x = \boxed{-3 \text{ or } 1}$.

146. The absolute values in the equation turn around when $x = 0$ or $7 - 2x = 0$ (i.e., $x = \frac{7}{2}$), so there are three cases to consider.
Case 1: $x \le 0$. The equation becomes $-x = 7 - 2x + 3$, so $x = 10$; this does not satisfy $x \le 0$, so there are no solutions in this region.
Case 2: $0 \le x \le \frac{7}{2}$. The equation becomes $x = 7 - 2x + 3$, so $x = \frac{10}{3}$, which is in the right range.
Case 3: $x \ge \frac{7}{2}$. The equation becomes $x = 2x - 7 + 3$, so $x = 4$, which is in the right range.
Thus the solutions are $x = \boxed{\frac{10}{3} \text{ or } 4}$.

147. **Solution 1:** If we look for an integral solution, we can quickly see that $21 = 3 \cdot 7$, $35 = 5 \cdot 7$, and $15 = 3 \cdot 5$, so $(a, b, c) = (3, 7, 5)$ and $a + b + c = \boxed{15}$.
Solution 2: Multiply all three equations together to get $a^2 b^2 c^2 = 3 \cdot 7 \cdot 3 \cdot 5 \cdot 5 \cdot 7 = 3^2 5^2 7^2$, so $abc = 3 \cdot 5 \cdot 7$. Dividing by each of the original equations in turn then gives $(a, b, c) = (3, 7, 5)$, so $a + b + c = \boxed{15}$.

148. Suppose that the first student takes a steps and the second b before their steps first coincide (after the very beginning). Then a steps of 60 cm cover the same distance as b steps of 69 cm, so $60a = 69b$, or $20a = 23b$. Since this is the first coincidence, a and b must be minimal; since 20 and 23 are relatively prime, we must have $x = 23$ and $y = 20$. Thus the students travel $23 \cdot 60$ cm $= 13.8$ m in 15 seconds; 5 minutes is 20 times as long, so the distance from A to B is $20 \cdot 13.8$ m $= \boxed{276 \text{ m}}$.

149. Cross multiplying and using combinations, we get

$$1990 \frac{(2n)!}{n!n!} = \frac{(2n+2)!(n+1)}{(n+1)!(n+1)!} = \frac{(2n+2)!(n+1)}{(n+1)n!(n+1)n!}$$
$$1990 \frac{(2n)!}{n!n!} = \frac{(2n+2)!(n+1)}{(n+1)n!(n+1)n!}.$$

Multiplying both sides of the above equation by $n!n!$ and simplifying, we get

$$1990(2n)!(n+1) = (2n+2)!$$
$$1990(n+1) = (2n+1)(2n+2)$$
$$1990 = 2(2n+1)$$
$$n = \boxed{497}.$$

Level 3

150. We can rearrange the given equation into

$$\frac{\sqrt{a}-\sqrt{b}}{2} = \frac{\sqrt{5}-\sqrt{3}+\sqrt{2}}{\sqrt{5}+\sqrt{3}+\sqrt{2}}$$
$$= \frac{\sqrt{5}-\sqrt{3}+\sqrt{2}}{\sqrt{5}+\sqrt{3}+\sqrt{2}} \cdot \frac{\sqrt{5}+\sqrt{3}-\sqrt{2}}{\sqrt{5}+\sqrt{3}-\sqrt{2}}$$
$$= \frac{5-(\sqrt{3}-\sqrt{2})^2}{(\sqrt{5}+\sqrt{3})^2-2}$$
$$= \frac{5-(5-2\sqrt{6})}{(8+2\sqrt{15})-2}$$
$$= \frac{2\sqrt{6}}{6+2\sqrt{15}} \cdot \frac{6-2\sqrt{15}}{6-2\sqrt{15}}$$
$$= \frac{12\sqrt{6}-4\sqrt{90}}{36-60}$$
$$= \frac{\sqrt{10}-\sqrt{6}}{2}.$$

Thus $(a,b) = \boxed{(10,6)}$.

151. We can rewrite the given equation as

$$\sqrt{3(x+1)^2+4} + \sqrt{5(x+1)^2+9} = 5-(x+1)^2.$$

The right side is at most 5 and, since $\sqrt{3(x+1)^2+4} \geq \sqrt{4} = 2$ and $\sqrt{5(x+1)^2+9} \geq 9 = 3$, the left side is at least 5. Thus they most both equal 5, which happens when $(x+1)^2 = 0$, so $x = \boxed{-1}$.

152. Factor both sides of each equation:

$$(a + c)(a + b) = 2 \cdot 7 \cdot 13,$$
$$(b + c)(b + a) = 2 \cdot 3 \cdot 7^2,$$
$$(c + a)(c + b) = 3 \cdot 7 \cdot 13.$$

Multiply all three together and take the positive square root:

$$(a + b)(b + c)(c + a) = 2 \cdot 3 \cdot 7^2 \cdot 13.$$

Divide by each of the original equations in turn:

$$b + c = 3 \cdot 7 = 21,$$
$$a + c = 13,$$
$$a + b = 2 \cdot 7 = 14.$$

Add these together and divide by 2 to get

$$a + b + c = 24,$$

and finally subtract each of the previous three equations in turn to get $(a, b, c) = \boxed{(3, 11, 10)}$.

153. We have $217 = x^3 - y^3 = (x - y)(x^2 + xy + y^2)$. Note that $x^3 > y^3$, so $x - y > 0$, and the only ways to factor 217 are $1 \cdot 217$ and $7 \cdot 31$.

If $x - y = 1$, then we need $217 = x^2 + xy + y^2 = (y^2 + 2y + 1) + (y^2 + y) + y^2 = 3y^2 + 3y + 1$, so $y(y + 1) = 72$, which is satisfied by $y = -9$ and $y = 8$.

If $x - y = 7$, then we need $31 = x^2 + xy + y^2 = (y^2 + 14y + 49) + (y^2 + 7y) + y^2 = 3y^2 + 21y + 49$, so we need $y(y + 7) = -6$, which is satisfied by $y = -1$ and $y = 6$.

Computing the corresponding value of x for each y gives the solutions $\boxed{(-8, -9), (9, 8), (6, -1), \text{ and } (1, -6)}$.

154. By definition, we have

$$s \leq x, \quad s \leq y + \frac{1}{x}, \quad s \leq \frac{1}{y}.$$

Since all the values are positive, inverting the first and third gives $\frac{1}{s} \geq \frac{1}{x}$ and $\frac{1}{s} \geq y$, and adding these gives $\frac{2}{s} \geq y + \frac{1}{x}$. Combining with the second gives $\frac{2}{s} \geq s$, or $s^2 \leq 2$, so $s \leq \sqrt{2}$. Equality is actually achieved when $(x, y) = \left(\sqrt{2}, \frac{1}{\sqrt{2}} \right)$, so the maximum is $\boxed{\sqrt{2}}$.

155. The nth term is equal to $\frac{3^n+1}{9^n}$, for $n \geq 0$, so we have

$$\sum_{n=0}^{\infty} \frac{3^n+1}{9^n} = \sum_{n=0}^{\infty} \frac{1}{3^n} + \frac{1}{9^n} = \frac{1}{1-\frac{1}{3}} + \frac{1}{1-\frac{1}{9}} = \frac{3}{2} + \frac{9}{8} = \boxed{\frac{21}{8}}.$$

156. Let the roots be $r_1 = \frac{a}{r}$, $r_2 = a$, and $r_3 = ar$; then Vieta's formulas give

$$-\frac{T}{19} = r_1 r_2 r_3 = a^3$$

$$\frac{23}{19} = r_1 r_2 + r_1 r_3 + r_2 r_3 = a^2 \cdot \left(\frac{1}{r} + 1 + r\right)$$

$$-\frac{92}{19} = r_1 + r_2 + r_3 = a \cdot \left(\frac{1}{r} + 1 + r\right).$$

Dividing the second equation by the third gives $a = -\frac{1}{4}$, and the first equation then gives $T = -19a^3 = \boxed{\frac{19}{64}}$.

157. **Solution 1:** Let $n = 1987$, so that

$$a = \sqrt{n-1} + \sqrt{n+1} \qquad a^2 = 2n + 2\sqrt{n^2-1}$$
$$b = 2\sqrt[4]{(n-1)(n+1)} \qquad b^2 = 4\sqrt{n^2-1}$$
$$c = 2\sqrt{n} \qquad c^2 = 4n.$$

Since $\sqrt{n^2-1} < n$, this means $b^2 < a^2 < c^2$, so, since they are all positive, in increasing order we have $\boxed{b, a, c}$.

Solution 2: Let $n = 1987$ and express a, b, and c as above; from the inequality of arithmetic means and geometric means, we have

$$\frac{a}{2} = \frac{\sqrt{n-1} + \sqrt{n+1}}{2} > \sqrt{\sqrt{n-1}\sqrt{n+1}} = \sqrt[4]{(n-1)(n+1)} = \frac{b}{2},$$

so $a > b$. Also, since the square root function is concave-down,

$$\frac{a}{2} = \frac{\sqrt{n-1} + \sqrt{n+1}}{2} < \sqrt{\frac{(n-1)+(n+1)}{2}} = \sqrt{n} = \frac{c}{2},$$

so $a < c$. Thus, in increasing order, we have $\boxed{b, a, c}$.

158. Let $a = 100$ and $b = a^2 + 3a = 10300$; then we have

$$\sqrt{(a(a+3))((a+1)(a+2)) + 1} = \sqrt{(a^2+3a)(a^2+3a+2) + 1}$$
$$= \sqrt{b(b+2) + 1}$$
$$= \sqrt{b^2 + 2b + 1}$$
$$= b + 1$$
$$= \boxed{10301}.$$

159. For any n, let $P_n = x^n + \frac{1}{x^n}$. Then $P_0 = 2$, $P_1 = a$, and we want to find P_5. In general,

$$a \cdot P_n = \left(x + \frac{1}{x}\right)\left(x^n + \frac{1}{x^n}\right)$$

$$= x^{n+1} + \frac{1}{x^{n-1}} + x^{n-1} + \frac{1}{x^{n+1}}$$

$$= P_{n+1} + P_{n-1}$$

$$P_{n+1} = a \cdot P_n - P_{n-1}.$$

Thus

$$P_0 = 2,$$
$$P_1 = a,$$
$$P_2 = a^2 - 2,$$
$$P_3 = a^3 - 3a,$$
$$P_4 = a^4 - 4a^2 + 2,$$
$$P_5 = \boxed{a^5 - 5a^3 + 5a}.$$

160. $x = \sqrt[3]{6 + \sqrt[3]{6 + \sqrt[3]{6 + \cdots}}}$ so $x = \sqrt[3]{6 + x}$. Cube both sides to get $x^3 = x + 6$ so $x^3 - x - 6 = 0$ and $(x - 2)(x^2 + 2x + 3) = 0$. So $x = \boxed{2}$. The other roots are complex.

161. Let $a = (1 + x)^{\frac{1}{7}}$ and $b = (1 - x)^{\frac{1}{7}}$, so the given equation becomes

$$a^2 + 3b^2 = 4ab$$
$$a^2 - 4ab + 3b^2 = 0$$
$$(a - b)(a - 3b) = 0,$$

and so either $a = b$ or $a = 3b$. The first implies $x = 0$, which we do not want. The second implies $(1 + x)^{\frac{1}{7}} = 3(1 - x)^{\frac{1}{7}}$, or $1 + x = 3^7(1 - x) = 2187(1 - x)$, so we have $2188x = 2186$ and $1094x = \boxed{1093}$.

162. Since x and y are positive, we can take the logarithms of the two equations, using base x and y respectively, to get

$$x + y = 9z \log_x y$$
$$x + y = z \log_y x.$$

Remember that, $\log_x y = \frac{1}{\log_y x}$; also, since $x + y > 0$, we know $z \neq 0$, so we can divide by z. Thus equating the right sides gives

$$(\log_y x)^2 = 9,$$
$$\log_y x = \pm 3,$$

so either $x = y^3$ or $x = y^{-3}$. In the first case, we have $16 = xy = y^4$, so $y = 2$ and $x = 8$; in the second, we have $16 = xy = y^{-2}$, so $y = \frac{1}{4}$ and $x = 64$. Thus the solutions are $\boxed{(8, 2) \text{ and } (64, \frac{1}{4})}$.

Chapter 3

Geometry

Questions

Level 1

1. Find the number of diagonals that can be drawn in a convex polygon with 40 sides.

2. A regular polygon has 54 distinct interior diagonals. How many sides does the polygon have?

3. Compute the length of the diagonal of an isosceles trapezoid with sides 5, 7, 7, and 13.

4. In quadrilateral $ABCD$, $\overline{AD} \perp \overline{CD}$ and $\overline{AB} \perp \overline{CB}$. If $BC = 1$, $CD = 2$, and $AD = 3$, compute AB.

5. Compute the radius of a circle in which a $40°$ arc measures 3π units in length.

6. Diagonal $\overline{AC}$ of square $ABCD$ is a side of square $ACEF$. If the area of $ACEF$ is 8, compute the area of $ABCD$.

7. The legs of a right triangle are 4 and 6. Find the length of the altitude to the hypotenuse of the triangle.

8. The sum of all but one of the interior angles of a convex polygon is $1930°$. Compute how many sides the polygon has.

9. In $\triangle ABC$, $BC = 10$. Points X and Y are on $\overline{AB}$ and $\overline{AC}$ respectively, and $\overline{XY} \| \overline{BC}$. If the area of $\triangle AXY$ is half the area of $\triangle ABC$, compute XY.

10. In $\triangle ABC$, $m\angle A = 70°$ and $m\angle B = 50°$. CH and CI are an altitude and an angle bisector of $\triangle ABC$. Compute $m\angle ICH$.

11. In a regular nine-sided polygon $A_1 A_2 A_3 A_4 \ldots A_9$, compute $m\angle A_1 A_9 A_4$.

12. Compute the area of a regular hexagon with side length 8.

13. $\triangle ABC$ has vertices $A(0,0)$, $B(4,0)$, and $C(0,3)$. Compute the length of the altitude from A to $\overline{BC}$.

14. Compute the radius of a sphere if its volume is numerically equal to its surface area.

15. The length of the tangent to a circle from an external point P is 7. If the radius of the circle is 3, compute the shortest distance from P to any point on the circle.

16. In $\triangle ABC$, $AB = 4$ and $BC = 7$. Find the ratio of the length of the altitude to $\overline{AB}$ to the length of the altitude to $\overline{BC}$.

17. Find the area of a regular octagon inscribed in a circle of radius 10.

18. $\triangle ABC$ is inscribed in circle O. The arc AB measures $130°$ and the arc BC measures $80°$. Point E is chosen on minor arc AC so that $\overline{OE} \perp \overline{AC}$. Find $m\angle OBE$.

19. A regular octagon is formed by cutting congruent isosceles right triangles from the corners of a square with side length 4. Find the length of a leg of one of these triangles.

20. $\overline{AB}$, $\overline{BC}$, and $\overline{CD}$ are consecutive sides of a regular polygon with n sides, where $n > 4$. $\overline{AB}$ and $\overline{CD}$ are extended through B and C respectively and meet at Q. Find $m\angle BQC$ in terms of n.

21. In parallelogram $ABCD$, $AB = 10$. Point P is the trisection point of $\overline{BD}$ closer to D and $\overline{AP}$ intersects $\overline{CD}$ at Q. Compute DQ.

Level 2

22. The sides of a triangle have lengths 10, 24, and 26. Compute the radius of the circle inscribed in this triangle.

23. In $\triangle ABC$, $AB = BC = 1$ and $m\angle B = 36°$. Compute AC.

24. In $\triangle ABC$, $m\angle B = 90°$, $AB = 1$, $BC = \sqrt{3}$, D is the midpoint of $\overline{AC}$, and P is on $\overline{BD}$ with $\overline{AP} \perp \overline{BD}$. Find AP.

25. In convex pentagon $ABCDE$, $AB = 1$, $BC = 2$, $CD = 3$, $DE = 4$, and $m\angle B = m\angle C = m\angle D = 135°$. If $x = AE$, compute the ordered pair of integers (p, q) such that $x^2 = p + q\sqrt{2}$.

26. Let A and B be the points $(21, 0)$ and $(0, 20)$. Let the point P be (a, b). If $\angle APB$ is a right angle, compute the minimum value that a can have.

27. Each side of an equilateral triangle has a semicircle constructed exterior to the triangle with the side as a diameter. The area of this figure (consisting of the triangle and three half-disks) is numerically

equal to its perimeter. Find the ordered pair of integers (a, b) such that the side length of the triangle is $\frac{a\pi^2 + b\pi\sqrt{3}}{3\pi^2 - 4}$.

28. M and N are points inside $\triangle ABC$ such that $m\angle CAM = m\angle BAN$, $m\angle CBN = m\angle ABM$, and $m\angle AMB + m\angle ANB = 200°$. Compute $m\angle ACB$.

29. $\triangle ABC$ is inscribed in a circle with $AB = 6$, $BC = 11$, and $CA = 7$. For all points P on minor arc $\overarc{AB}$, $CP = m \cdot AP + n \cdot BP$. Compute $m + n$.

30. If a scalene triangle has area $3\sqrt{15}$ and two of its medians have lengths 3 and 6, compute the length of the third median.

31. Two chords of a circle are perpendicular. One has segments of 4 and 3 and the other has segments of 2 and 6. Find the radius of the circle.

32. In a circle of radius 10, two parallel chords are drawn on opposite sides of the center, each 5 units from the center. Find the area of the region contained between the chords and within the circle.

33. Compute the area of hexagon $ABCDEF$, which is formed by connecting the midpoints of adjacent edges of a unit cube as shown in Fig. 3.1.

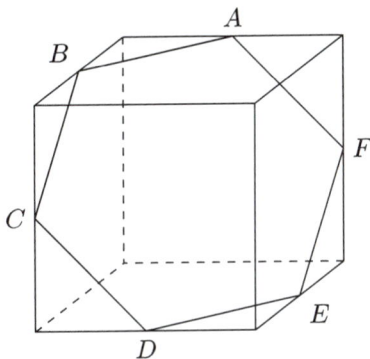

Figure 3.1

34. Find the volume of the pyramid $PEFGH$, as shown in Fig. 3.2, where $EFGH$ is a face of unit cube $ABCDEFGH$ and P is the intersection of $\overline{AG}$ and $\overline{FD}$.

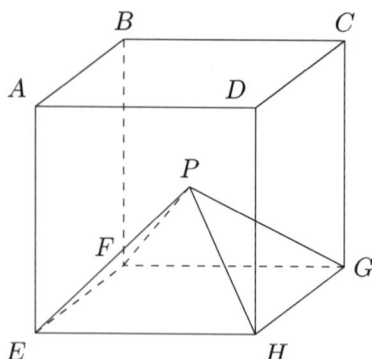

Figure 3.2

35. Parallelogram $ABCD$ is inscribed in a circle. If $AB = 4$ and $BC = 6$, compute the radius of the circle.

36. Two congruent 30–60–90 triangles with hypotenuse 6 are placed so that their hypotenuses coincide and they overlap in a region which has nonzero area but is not all of the triangles. Compute the area of the overlap.

37. In $\triangle ABC$, $AC = 5$, $CB = 12$, and $AB = 13$. The bisector of $\angle CAB$ intersects $\overline{CB}$ at D. Compute CD.

38. In pentagon $ABCDE$, $\overline{AE} \perp \overline{AB}$ and $\overline{BC} \perp \overline{CD}$. F is the foot of the perpendicular from B to $\overline{DE}$. If $AB = 3$, $BC = 4$, $CD = 5$, $DF = 6$, and $EF = 7$, compute AE.

39. In a circle, chord $AB = 24$. Point M is the midpoint of minor arc $\overset{\frown}{AB}$, and chord $AM = 13$. Compute the radius of the circle.

40. From point P outside circle O, tangents are drawn from P to points X and Y on O. If $PO = PX + PY$, compute $m\angle XPY$.

41. Compute the radius of the circle centered at $(0,0)$ and tangent to the line described by $x + 2y = 10$.

42. The lengths of the sides of a rectangle are all integers, the number of square units in its area is one less than the number of linear units in its perimeter. Compute the area of the rectangle.

43. In $\triangle ABC$, $AB = AC$, point D is on $\overline{AC}$ and $AD = DB = BC$. Compute $m\angle A$.

44. In square $ABCD$, point M is the midpoint of $\overline{BC}$, and point N is on $\overline{CD}$ such that $\overline{MN} \perp \overline{AM}$. Compute the ratio $CN{:}ND$.

45. Circle O is inscribed in $\triangle ABC$, and $AB = AC$. Points M and N are midpoints of $\overline{AB}$ and $\overline{AC}$ respectively, and $\overline{MN}$ is tangent to circle O. Compute the ratio $AB{:}BC$.

46. Two circles are concentric and a chord of the larger circle is tangent to the smaller circle. If this chord is 12 units long, compute the area of the region between the circles.

47. In $\triangle ABC$, $AB = AC$. There are points D on $\overline{AB}$, E on $\overline{CA}$, and F on $\overline{AD}$ such that $CB = CD = ED = EF = FA$. Compute $m\angle A$.

48. In $\triangle ABC$, the line containing the bisector of angle B and the bisector of an exterior angle at C intersect at Q. A line through Q parallel to $\overline{BC}$ intersects $\overline{AC}$ at E and $\overline{AB}$ at D. If $BD = 8$ and $EC = 6$, compute ED.

49. In $\triangle ABC$, $AB = AC = 13$ and $BC = 10$. Point P is on $\overline{BC}$. If X and Y are the feet of the perpendiculars from P to $\overline{AB}$ and $\overline{AC}$ respectively, compute $PX + PY$.

50. In $\triangle ABC$, $AB = AC = 17$. Point E is the trisection point of $\overline{BC}$ nearer B and $AE = 15$. Compute BC.

51. In trapezoid $ABCD$, M and N are the midpoints of legs AD and BC respectively. $\overline{MN}$ intersects $\overline{AC}$ at P and $\overline{BD}$ at Q. If $AB = 7$ and $CD = 13$, compute PQ.

52. Circles O_1 and O_2 have radii 16 and 9 respectively and are externally tangent at point A. Point B is chosen on circle O_1 such that $AB = 8$ and $\overline{BC}$ is drawn tangent to circle O_2 at point C. Compute BC.

53. In right triangle ABC, D is a point on leg $\overline{BC}$, M is the midpoint of hypotenuse $\overline{AB}$, and $\overline{MD} \perp \overline{AB}$. If $AC = 8$ and $DC = 6$, compute AB.

54. In a rectangular coordinate system, there are two circles passing through $(3, 2)$ and tangent to both coordinate axes. Compute the sum of the radii of the circles.

55. Squares $ABCD$ and $ABEF$ are adjacent sides of a cube. Compute $m\angle FBD$.

56. A leg of an isosceles trapezoid is 10 units long and a circle inscribed in the trapezoid has radius 2 units. Compute the area of the trapezoid.

57. Squares $ABCD$ and $ABQP$ are adjacent faces of a cube. Point E is the midpoint of $\overline{QP}$. The plane containing A, C, and E intersects a fourth edge of the cube at F. Compute the area of the quadrilateral $ACFE$, if $AB = 4$.

58. The centers of two externally tangent circles are 7 units apart. If the length of a common external tangent to both circles is 6 units, compute the positive difference between their radii.

59. In rhombus $PQRS$, points A and B are on $\overline{QR}$ and $\overline{RS}$ respectively such that $\triangle PAB$ is equilateral. If $PQ = PA$, compute $m\angle PQR$.

60. Compute the distance between two opposite vertices of a cube whose edge length is 4 units.

61. Compute the edge length of a cube which is inscribed in a sphere of radius 6 units.

62. Acute triangle $\triangle A_1 B_1 C_1$ is inscribed in a circle. Points A_2, B_2, and C_2 are the midpoints of minor arcs $\overset{\frown}{B_1 C_1}$, $\overset{\frown}{A_1 C_1}$, and $\overset{\frown}{A_1 B_1}$ respectively. Points A_3, B_3, and C_3 are midpoints of minor arcs $\overset{\frown}{B_2 C_2}$, $\overset{\frown}{A_2 C_2}$, and $\overset{\frown}{A_2 B_2}$ respectively. If $m\angle A_3 B_3 C_3 = m\angle A_1 B_1 C_1$, compute $m\angle A_1 B_1 C_1$.

63. In square $ABCD$, $AB = 2$. Four line segments of length 2 are constructed, each with one endpoint at a different vertex of the square. These four segments have a common endpoint outside the plane of the square, at point P. Compute $m\angle APB$.

64. The sum of the lengths of the diagonals of a rhombus is 14 units and its area is 13 square units. Compute the side length of the rhombus.

Level 3

65. In $\triangle ABC$, D is the midpoint of $\overline{AB}$ and E is the midpoint of $\overline{BD}$. Point F is on $\overline{AB}$ such that E is on $\overline{CF}$. If $BF = 5$, compute BA.

66. X is outside $\triangle ABC$ and P, Q, and R are respectively on $\overline{AB}$, $\overline{BC}$, and $\overline{AC}$ such that $\overline{XP} \perp \overline{AB}$, $\overline{XQ} \perp \overline{BC}$, and $\overline{XR} \perp \overline{AC}$. If $BP = 1$, $BQ = 2$, $CQ = 3$, $CR = 4$, and $AR = 5$, compute AP.

67. In rhombus $ABCD$, the radius of the circle through points A, B, and C is 6, while the radius of the circle through points A, B, and D is 2. Compute the area of $ABCD$.

68. In $\triangle KLM$, the angle bisectors $\overline{KN}$ and $\overline{LP}$ intersect at Q, and point M lies on the circle through P, Q, and N. If $PN = 2$, compute PQ.

69. A quadrilateral $ABCD$ is split into two triangles of equal area by diagonal $\overline{AC}$. If A, B, and C are $(1,1)$, $(17,48)$, and $(7,7)$ respectively, and D is (a,b), compute $a - b$.

70. In equilateral triangle $\triangle ABC$, $AB = 15$. Point D is the trisection point of $\overline{BC}$ closer to B, and point E is on $\overline{AB}$ and equidistant from A and D. Compute CE.

71. A paper rectangle of size 10×24 units is folded and creased so that two opposite vertices coincide. Compute the length of the crease formed.

72. Hexagon $ABCDEF$ is inscribed in a circle with $AB = CD = EF = 2$ and $BC = DE = FA = 10$. Compute the area of an equilateral triangle inscribed in the same circle.

Answers

Level 1

1. The number of diagonals of a polygon with n sides is $\frac{n(n-3)}{2}$. In this case, we have $\frac{40 \cdot 37}{2} = \boxed{740}$.

2. We have $\frac{n(n-3)}{2} = 54$, or $n^2 - 3n - 108 = 0$. This factors into $(n+9)(n-12) = 0$, which has a positive root of $\boxed{12}$.

3. Draw altitude $\overline{AE}$. Since the trapezoid is isosceles, $ED = \frac{CD-AB}{2} = 4$. We have $AE^2 = AD^2 - ED^2 = 33$ by the Pythagorean theorem in $\triangle AED$. Then, by the Pythagorean theorem in $\triangle AEC$ (see Fig. 3.3), we have $AC = \sqrt{CE^2 + AE^2} = \sqrt{81 + 33} = \boxed{\sqrt{114}}$.

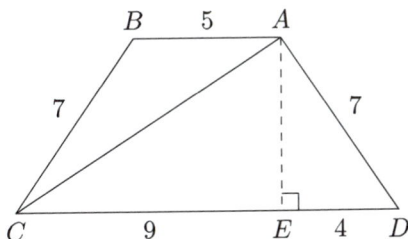

Figure 3.3

4. Applying the Pythagorean theorem to $\triangle ABC$ and $\triangle ADC$, which both have $\overline{AC}$ as a hypotenuse, gives $AB^2 + 1^2 = AC^2 = AD^2 + CD^2$, or $AB^2 + 1 = 13$, so $AB = \sqrt{12} = \boxed{2\sqrt{3}}$ (see Fig. 3.4).

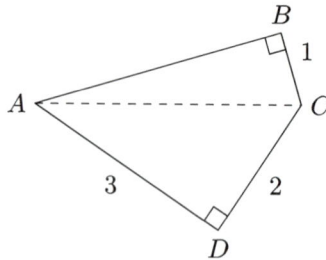

Figure 3.4

5. A $40°$ arc is $\frac{1}{9}$ of a circle or $\frac{2\pi}{9}$ radians, so $\frac{2\pi}{9} r = 3\pi$ and $r = \boxed{\frac{27}{2}}$.

6. $AC = \sqrt{8}$ and $AB = \frac{AC}{\sqrt{2}} = 2$, so the area of $ABCD$ is $2^2 = \boxed{4}$. Alternately, the area of a square is half the product of its diagonals, which is $\frac{\sqrt{8} \cdot \sqrt{8}}{2} = 4$.

7. The area of the triangle is half the product of its legs, which is $\frac{4 \cdot 6}{2} = 12$, and its hypotenuse is $\sqrt{4^2 + 6^2} = \sqrt{52}$. If h is the length of the altitude to the hypotenuse, then the area of the triangle is also $\frac{\sqrt{52} \cdot h}{2} = h\sqrt{13}$. Thus $h = \frac{12}{\sqrt{13}} = \boxed{\frac{12\sqrt{13}}{13}}$.

8. The sum of the measures of the interior angles of a convex polygon with n sides is $(n-2) \cdot 180°$. Since any individual angle of a convex polygon is between $0°$ and $180°$, the sum of all the angles of this polygon must be the smallest multiple of $180°$ greater than $1930°$. This multiple is $1980° = 11 \cdot 180°$, so $n = 11 + 2 = \boxed{13}$.

9. The ratio of the areas of similar triangles is equal to the square of the ratio of their sides. We are given that the ratio of the areas of similar triangles $\triangle AXY$ and $\triangle ABC$ is 1:2, so $XY{:}BC = 1{:}\sqrt{2}$, so $XY = \frac{BC}{\sqrt{2}} = \boxed{5\sqrt{2}}$.

10. $m\angle C = 180° - m\angle A - m\angle C = 60°$ and $m\angle ICA = \frac{m\angle C}{2} = 30°$. From right triangle $\triangle ACH$ (see Fig. 3.5), we have $m\angle HCA = 90° - m\angle A = 20°$. Finally, $m\angle ICH = m\angle ICA - m\angle HCA = \boxed{10°}$.

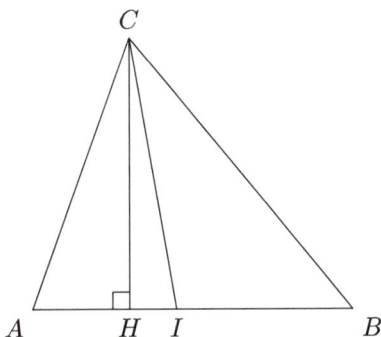

Figure 3.5

11. In the circumscribing circle of the polygon, the arc cut off by any one side measures $\frac{360°}{9} = 40°$. Hence the measure of arc $\overset{\frown}{A_1 A_4}$ is $3 \cdot 40° = 120°$, and so the measure of the inscribed angle $\angle A_1 A_9 A_4$ is $\frac{120°}{2} = \boxed{60°}$.

12. If we draw segments from the center of the hexagon to the vertices, we divide the hexagon into six equilateral triangles with side length 8. Since the area of an equilateral triangle with side length s is $\frac{s^2\sqrt{3}}{4}$, the total area is $6 \cdot \frac{8^2\sqrt{3}}{4} = \boxed{96\sqrt{3}}$.

13. $\triangle ABC$ is a $3-4-5$ right triangle; its area is half the product of its legs, which is $\frac{3 \cdot 4}{2} = 6$. If h is the length of the altitude to the hypotenuse, then the area of the triangle is also $\frac{5 \cdot h}{2}$. Thus $h = 6 \cdot \frac{2}{5} = \boxed{\frac{12}{5}}$.

14. The volume of the sphere is $\frac{4}{3}\pi r^3$ cubic units and its surface area is $4\pi r^2$ square units, so we have $\frac{4}{3}\pi r^3 = 4\pi r^2$, so $r = \boxed{3}$.

15. A tangent to a circle is perpendicular to the radius to the tangent point and the closest point on the circle to P is the intersection of

$\overline{OP}$ with the circle (see Fig. 3.6). We have $(x+3)^2 = 7^3 + 3^2 = 58$, so $x = \boxed{\sqrt{58} - 3}$.

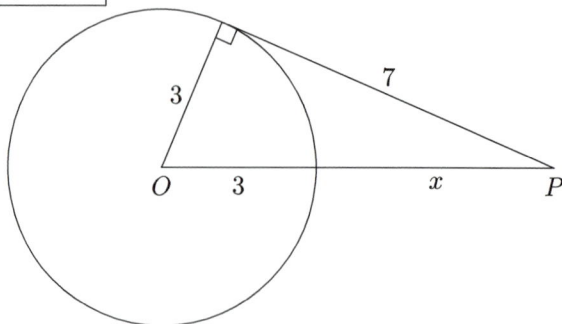

Figure 3.6

16. Let $a = BC$ and $c = AB$. Let k be the area of the triangle and h_a and h_c be the lengths of the altitudes to $\overline{BC}$ and $\overline{AB}$. Then $k = \frac{a \cdot h_a}{2} = \frac{c \cdot h_c}{2}$, so $a \cdot h_a = c \cdot h_c$ and $\frac{h_c}{h_a} = \frac{a}{c} = \boxed{\frac{7}{4}}$.

17. The octagon can be partitioned into eight triangles (see Fig. 3.7), each with area $\frac{10 \cdot 10 \cdot \sin 45^\circ}{2} = 25\sqrt{2}$, so the area of the octagon is $8 \cdot 25\sqrt{2} = \boxed{200\sqrt{2}}$.

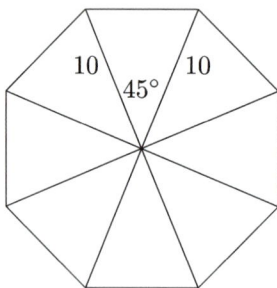

Figure 3.7

18. Since $\overline{OE}$ is perpendicular to $\overline{AC}$, it bisects arc $\widehat{AC}$, so arcs $\widehat{AE}$ and $\widehat{EC}$ both measure $\frac{360^\circ - 130^\circ - 80^\circ}{2} = 75^\circ$ (see Fig. 3.8). Thus

$m\angle BOE = 80° + 75° = 155°$; since $\triangle BOE$ is isosceles, $m\angle OBE = \frac{180° - 155°}{2} = \boxed{12.5°}$.

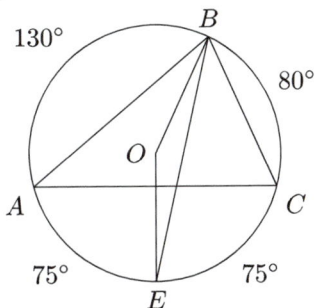

Figure 3.8

19. Let x be the leg of one of the triangles (see Fig. 3.9). Then the hypotenuse is $x\sqrt{2}$ and the remaining portions of an original side of the square is $4 - 2x$, so

$$4 - 2x = x\sqrt{2}$$
$$4 = x \cdot (\sqrt{2} + 2)$$
$$x = \frac{4}{\sqrt{2} + 2} = \frac{4(2 - \sqrt{2})}{2}$$
$$= \boxed{4 - 2\sqrt{2}}.$$

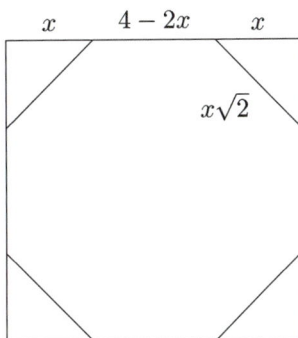

Figure 3.9

20. Since $\angle QBC$ and $\angle QCB$ are exterior angles of a regular polygon, they each have measure $\frac{360°}{n}$. Thus, $m\angle BQC = 180° - \frac{360°}{n} - \frac{360°}{n} = \boxed{180° - \frac{720°}{n}}$.

21. $\triangle APB \sim \triangle QPD$, so $\frac{x}{10} = \frac{y}{2y} = \frac{1}{2}$, so $x = \boxed{5}$ (see Fig. 3.10).

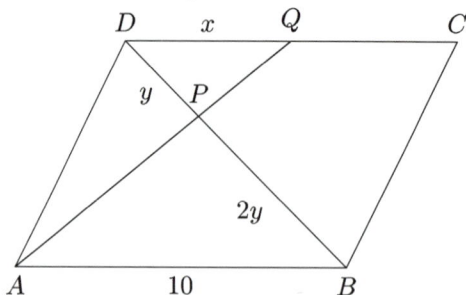

Figure 3.10

Level 2

22. Since $\triangle ABC$ is a right triangle, its area is $\frac{10 \cdot 24}{2} = 120$ (see Fig. 3.11). This is also the sum of the areas of $\triangle AOB$, $\triangle BOC$, and $\triangle COA$. So we have $120 = \frac{10 \cdot r + 24 \cdot r + 26 \cdot r}{2} = 30r$, so $r = \boxed{4}$. (Note that this is simply a specific case of the proof of the formula $k = rs$.)

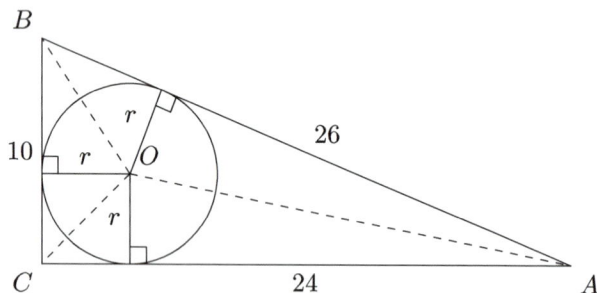

Figure 3.11

23. Note $m\angle A = m\angle C = \frac{180° - 36°}{2} = 72°$. Let $AC = x$ and draw the angle bisector $\overline{CD}$ (see Fig. 3.12). Then $m\angle ADC = 72° = m\angle A$, so $AC = CD = x$, and $m\angle BCD = 36° = m\angle B$, so $CD = DB = x$.

Finally, $\triangle ACD \sim \triangle CBA$, so $\frac{x}{1-x} = \frac{1}{x}$, or $x^2 + x - 1 = 0$, which has a positive root of $\boxed{\frac{-1+\sqrt{5}}{2}}$.

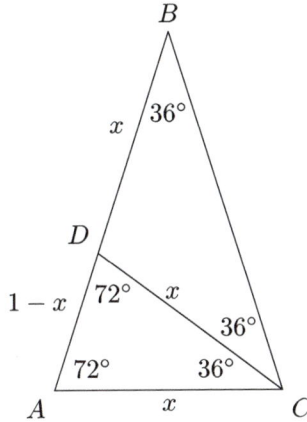

Figure 3.12

24. By the Pythagorean theorem, $AC = 2$ and so $AD = 1$ (see Fig. 3.13). Since the median to the hypotenuse of a right triangle is half as long as the hypotenuse, $BD = 1$ as well, so $\triangle ABD$ is equilateral. This makes $\triangle APB$ a $30-60-90$ triangle, so $AP = \boxed{\frac{\sqrt{3}}{2}}$.

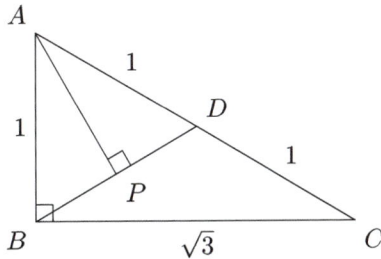

Figure 3.13

25. Position the polygon so that $\overline{CD}$ is horizontal, as in Fig. 3.14. Let F and G be the feet of the perpendiculars from B and E to $\overleftrightarrow{CD}$. Then $m\angle BCF = 180° - m\angle BCD = 45°$ and $m\angle EDG = 180° - m\angle EDC = 45°$. Thus both $\triangle BCF$ and $\triangle EDG$ are $45-45-90$

right triangles, so $BF = FC = \frac{BC}{\sqrt{2}} = \sqrt{2}$ and $EG = GD = \frac{ED}{\sqrt{2}} = 2\sqrt{2}$. Also, since $m\angle CBF = 45° = 180° - m\angle ABC$, the points A, B, and F are collinear, and so $\overline{AB}$ is vertical, since $\overline{BF}$ is, by definition, perpendicular to $\overline{CD}$. Then the horizontal distance between A and E is $FG = FC + CD + DG = 3 + 3\sqrt{2}$ and the vertical distance between them is $|AF - EG| = |(AB + BF) - EG| = \sqrt{2} - 1$. Thus, the square of the total distance between them is $(3 + 3\sqrt{2})^2 + (\sqrt{2} - 1)^2 = 30 + 16\sqrt{2}$, so the answer is $\boxed{(30, 16)}$.

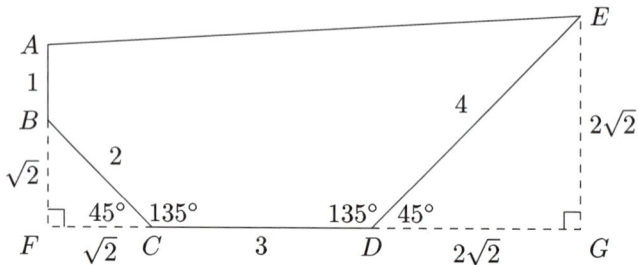

Figure 3.14

26. All possible locations of P lie on the circle with AB as a diameter. This circle is centered at the midpoint of $\overline{AB}$, which is $\left(\frac{21}{2}, 10\right)$, and has a radius of $\frac{AB}{2} = \frac{\sqrt{20^2 + 21^2}}{2} = \frac{29}{2}$ (see Fig. 3.15). Thus, the minimum x coordinate of any point on the circle is $\frac{21}{2} - \frac{29}{2} = \boxed{-4}$.

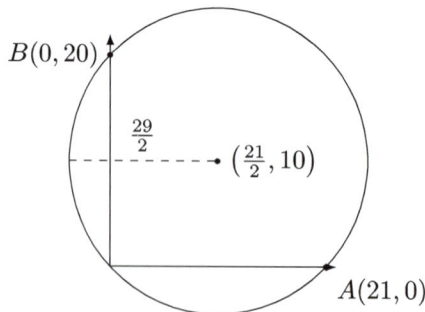

Figure 3.15

27. If the side of the triangle is s (see Fig. 3.16), each semicircle contributes $\frac{\pi}{2}\left(\frac{s}{2}\right)^2 = \frac{\pi}{8}s^2$ to the area and $\frac{\pi s}{2}$ to the circumference, and

the triangle itself contributes $\frac{s^2\sqrt{3}}{4}$ to the area. Thus, the area of the shape is $(\frac{\sqrt{3}}{4} + \frac{3\pi}{8})s^2$ and its perimeter is $\frac{3\pi s}{2}$, so

$$\left(\frac{\sqrt{3}}{4} + \frac{3\pi}{8}\right)s = \frac{3\pi}{2}$$

$$s = \frac{12\pi}{2\sqrt{3} + 3\pi}$$

$$= \frac{12\pi(3\pi - 2\sqrt{3})}{9\pi^2 - 12}$$

$$= \frac{12\pi^2 - 8\pi\sqrt{3}}{3\pi^2 - 4}.$$

Thus $(a, b) = \boxed{(12, -8)}$.

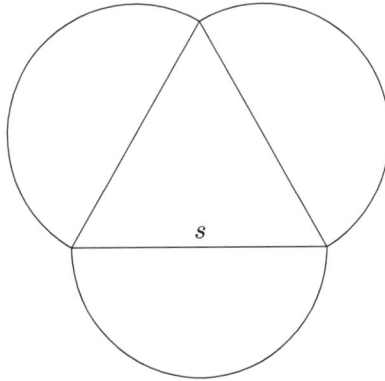

Figure 3.16

28. Note that $m\angle BAM = m\angle A - m\angle CAM = m\angle A - m\angle BAN$ and so $m\angle BAM + m\angle BAN = m\angle A$; similarly, $m\angle ABM + m\angle ABN = m\angle B$ (see Fig. 3.17). Adding up all the angles of both $\triangle ABM$ and $\triangle ABN$ gives

$$m\angle ABM + m\angle AMB + m\angle BAM$$
$$+ m\angle ABN + m\angle ANB + m\angle BAN = 360°$$

$$m\angle AMB + m\angle ANB + m\angle A + m\angle B = 360°$$
$$m\angle A + m\angle B = 160°.$$

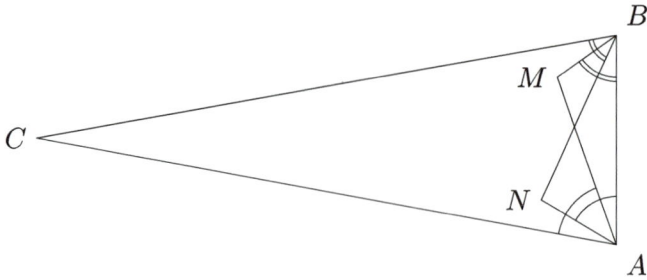

Figure 3.17

Thus, $m\angle C = 180° - m\angle A - m\angle B = \boxed{20°}$.

29. Since $APBC$ is cyclic, Ptolemy's theorem gives $AB \cdot CP = AP \cdot BC + AC \cdot BP$, or $6 \cdot CP = 11 \cdot AP + 7 \cdot BP$, so $CP = \frac{11}{6}AP + \frac{7}{6}BP$ and $m + n = \frac{11}{6} + \frac{7}{6} = \boxed{3}$ (see Fig. 3.18).

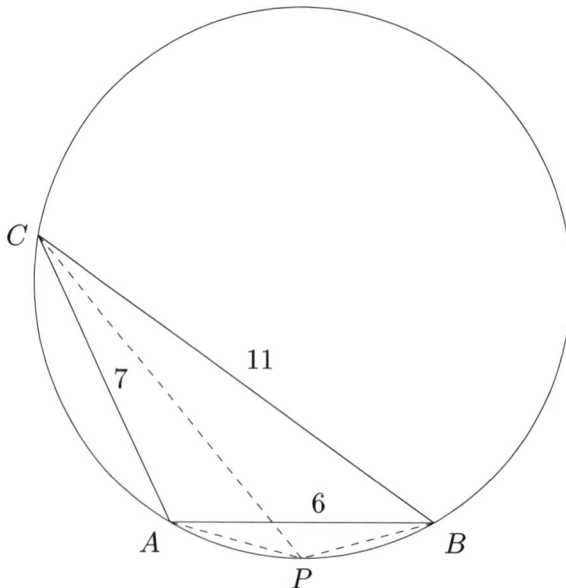

Figure 3.18

30. The area of a triangle the lengths of whose sides are the lengths of the medians of another triangle is $\frac{3}{4}$ times the area of the original triangle (see appendix). In the new triangle, let θ be the angle between the sides of length 3 and 6. Then $\frac{1}{2} \cdot 3 \cdot 6 \cdot \sin\theta = \frac{3}{4} \cdot 3\sqrt{15}$, or $\sin\theta = \frac{\sqrt{15}}{4}$, so $\cos\theta = \pm\sqrt{1 - (\frac{\sqrt{15}}{4})^2} = \pm\frac{1}{4}$. Let m be the length of the third median; by the law of cosines, $m^2 = 3^2 + 6^2 \pm 2 \cdot 3 \cdot 6 \cdot \frac{1}{4} = \{36, 54\}$. If $m^2 = 36$, then $m = 6$ and the original triangle is not scalene. (If two medians of a triangle are equal, then the triangle is isosceles; you can prove this using Stewart's theorem.) Thus $m = \sqrt{54} = \boxed{3\sqrt{6}}$. (Note that we don't need to check the value any further, because any triangle is the median triangle of some other triangle: take a triangle and translate all of its edges so that all of their clockwise vertices move to the same point. Then push each edge through the shared point by one third of its length. Then the edges are the medians of the triangle formed by the endpoints of the longer sections of the edges.)

31. The center of the circle (O in Fig. 3.19) is on the perpendicular bisectors of both chords, and the midpoints of the chords divide the chords into sections of the marked lengths. Thus the radius of the circle is $\sqrt{2^2 + 3.5^2} = \boxed{\frac{\sqrt{65}}{2}}$.

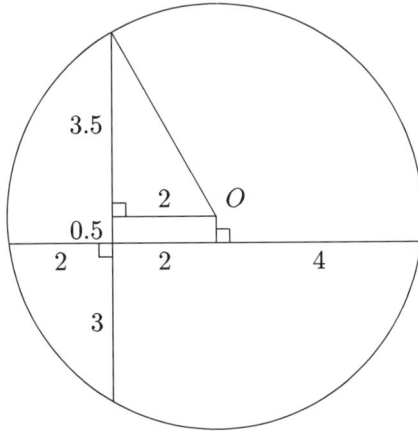

Figure 3.19

32. The region consists of two $60°$ sectors, each of area $\frac{1}{6} \cdot 100\pi = \frac{50}{3}\pi$ and two triangles, each of area $5 \cdot 5\sqrt{3} = 25\sqrt{3}$ (see Fig. 3.20). Thus its total area is $\boxed{\frac{100}{3}\pi + 50\sqrt{3}}$.

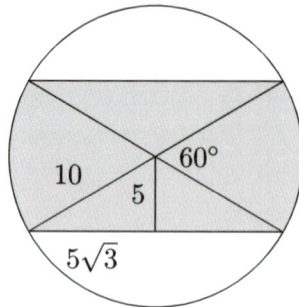

Figure 3.20

33. Each side of the hexagon has length $s = \frac{\sqrt{2}}{2}$; the hexagon consists of six equilateral triangles with side s, so its total area is $6 \cdot \frac{s^2\sqrt{3}}{4} = \boxed{\frac{3\sqrt{3}}{4}}$.

34. The intersection of the diagonals is the center of the cube, so the height of the pyramid is $\frac{1}{2}$. Its base has area 1, being a unit square, so the volume of the pyramid is $\frac{1}{3} \cdot \frac{1}{2} \cdot 1 = \boxed{\frac{1}{6}}$.

35. **Solution 1:** If parallelogram $ABCD$ (see Fig. 3.21) is inscribed in a circle, then arcs $\overarc{ABC}$ and $\overarc{CDA}$ are congruent, so each measures $180°$, which means that $m\angle B = 90°$; the same applies to every vertex, so the parallelogram is a rectangle, and a diagonal is a diameter of the circle. Thus the diagonals have length $\sqrt{6^2 + 4^2} = \sqrt{52}$, and the radius of the circle is $\frac{\sqrt{52}}{2} = \boxed{\sqrt{13}}$.

Solution 2: The center of the circle must be the center of the parallelogram (if it were otherwise, rotating by $180°$ around the center of the parallelogram would produce two distinct circles that intersected in four points). Thus the diagonals of the parallelogram are diameters; since every vertex is an inscribed angle of one of the diagonals, all the vertices are right angles. Thus the diagonals have length $\sqrt{6^2 + 4^2} = \sqrt{52}$, and the radius of the circle is $\frac{\sqrt{52}}{2} = \boxed{\sqrt{13}}$.

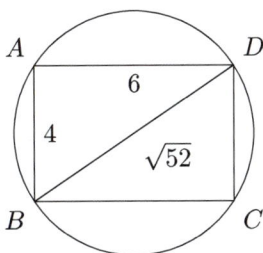

Figure 3.21

36. **Solution 1:** The two halves of the overlap region (as divided by the dashed line in Fig. 3.22) are both 30−60−90 triangles, so the height of the overlap is $\frac{3}{\sqrt{3}} = \sqrt{3}$ and its area is $\frac{1}{2} \cdot 6 \cdot \sqrt{3} = \boxed{3\sqrt{3}}$.

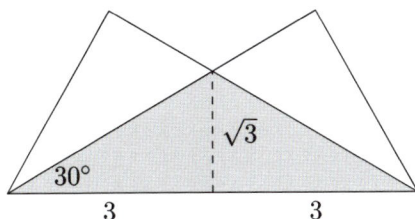

Figure 3.22

Solution 2: The overlap region is one-third of the large equilateral triangle shown in Fig. 3.23 (with the other two congruent thirds shown filled in with different patterns), so its area is $\frac{1}{3} \cdot \frac{6^2\sqrt{3}}{4} = \boxed{3\sqrt{3}}$.

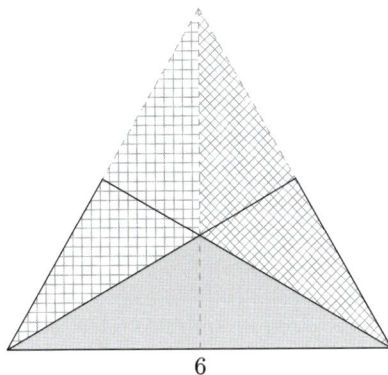

Figure 3.23

Solution 3: The four small triangles in Fig. 3.24 are all congruent; the overlap consists of two of them, while each original triangle consists of three. The legs of the original triangle are 3 and $3\sqrt{3}$, so the area of the overlap is $\frac{2}{3} \cdot \frac{1}{2} \cdot 3 \cdot 3\sqrt{3} = \boxed{3\sqrt{3}}$.

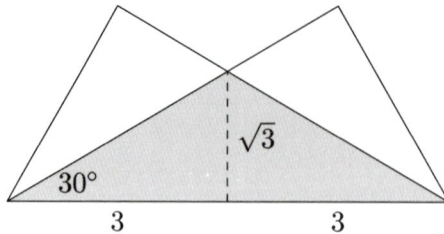

Figure 3.24

37. By the angle bisector theorem, $CD{:}DB = CA{:}CB = 5{:}13$, so we can write $CD = 5x$ and $DB = 13x$ (see Fig. 3.25). We know $12 = BC = CD + DB = 18x$, so $x = \frac{2}{3}$ and $CD = \boxed{\frac{10}{3}}$.

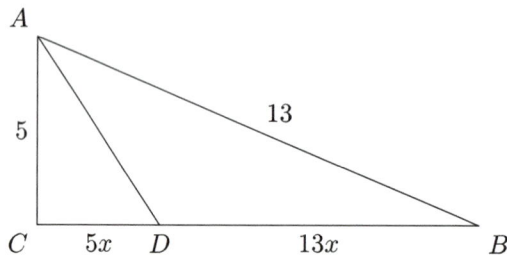

Figure 3.25

38. By repeated application of the Pythagorean theorem (see Fig. 3.26), we have:

$$BD^2 = BC^2 + CD^2 = 41$$
$$BF^2 = BD^2 - DF^2 = 5$$

$$BE^2 = BF^2 + EF^2 = 54$$
$$AE^2 = BE^2 - AB^2 = 45.$$

So $AE = \sqrt{45} = \boxed{3\sqrt{5}}$.

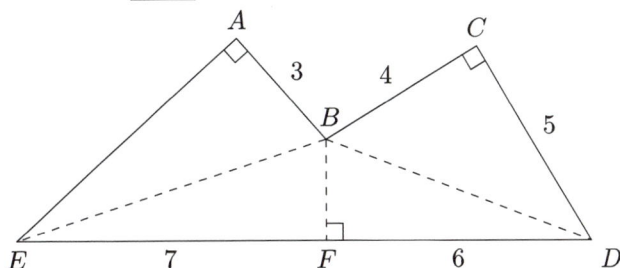

Figure 3.26

39. Since $\overline{AM}$ is perpendicular to and bisects $\overline{AB}$, $\triangle AMP$ is a $5-12-13$ right triangle (see Fig. 3.27). By the Chord-Chord theorem, $NP \cdot PM = AP \cdot PB$, or $5 \cdot NP = 12 \cdot 12$, so $NP = \frac{144}{5}$. The radius of the circle is then $\frac{NP+PM}{2} = \boxed{\frac{169}{10}}$.

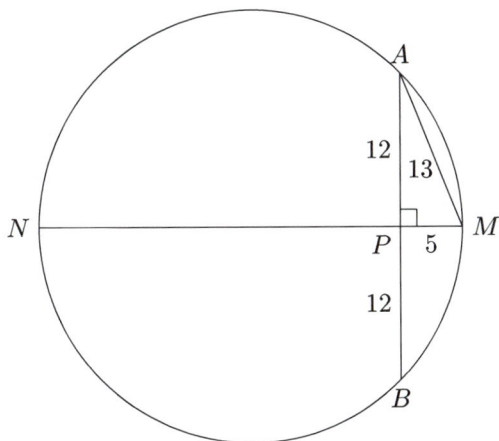

Figure 3.27

40. By symmetry, $PX = PY$ and so $PO = 2PX$ (see Fig. 3.28). Since $\overline{PX}$ is tangent to the circle, $\angle PXO$ is a right angle, and $\triangle PXO$ must be a $30-60-90$ right triangle. Thus $m\angle XPY = 2m\angle OPX = \boxed{120°}$.

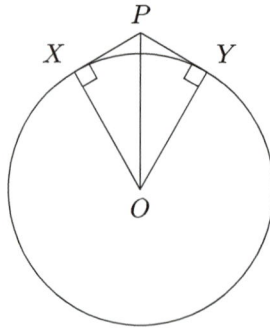

Figure 3.28

41. **Solution 1:** The radius is the distance from $(0,0)$ to the line $x + 2y - 10 = 0$; applying the general formula for the distance from a point to a line gives the distance as $\frac{|1 \cdot 0 + 2 \cdot 0 - 10|}{\sqrt{1^2 + 2^2}} = \boxed{2\sqrt{5}}$.

    Solution 2: The radius is the distance, as above, but we can also compute it more geometrically. By equating areas (see Fig. 3.29), we have $5 \cdot 10 = d \cdot 5\sqrt{5}$; by similar triangles, we have $\frac{5}{d} = \frac{5\sqrt{5}}{10}$ or $\frac{10}{d} = \frac{5\sqrt{5}}{5}$; all of these yield $d = \boxed{2\sqrt{5}}$.

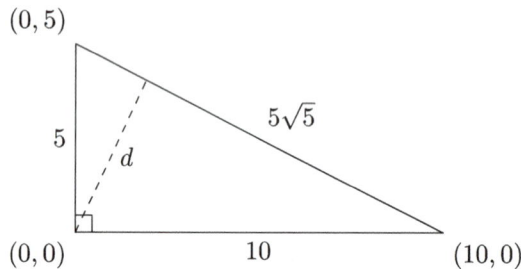

Figure 3.29

42. Let the dimensions of the rectangle be a and b. Then

$$ab + 1 = 2a + 2b,$$
$$ab - 2a - 2b + 4 = 3,$$
$$(a - 2)(b - 2) = 3.$$

Since a and b are integers, so are $a - 2$ and $b - 2$. So $a - 2$ and $b - 2$ must equal 1 and 3, meaning that a and b equal 3 and 5. (Of course, we also have $-1 \cdot -3 = 3$, but that would lead to negative dimensions.) Thus the area of the rectangle is $3 \cdot 5 = \boxed{15}$.

43. Let $x = m\angle A$. Because $\triangle ADB$ is isosceles (see Fig. 3.30), $m\angle ABD = m\angle BAD = x$. Since $\angle BDC$ is an exterior angle of $\triangle ADB$, $m\angle BDC = m\angle ABD + m\angle BAD = 2x$; since $\triangle CBD$ is isosceles, $m\angle CDB = m\angle DCB = 2x$. Since $\triangle ABC$ is isosceles, $m\angle CBA = m\angle BCA = 2x$. Finally, the sum of the angles of $\triangle ABC$ must equal $180°$, which gives $x + 2x + 2x = 180°$, so $x = \boxed{36°}$.

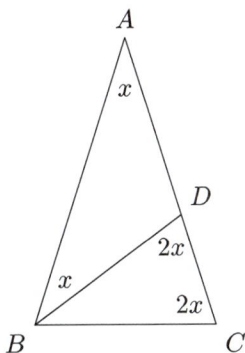

Figure 3.30

44. Let s be the side length of the square (see Fig. 3.31). We have $m\angle BAM + m\angle BMA = 90°$ and $m\angle BMA + m\angle AMN + m\angle NMC = 180°$, so $m\angle BAM = m\angle NMC$ and $\triangle BAM \sim \triangle CMN$, with ratio of similitude $AB{:}MC = 2{:}1$. Since $BM = \frac{s}{2}$, $CN = \frac{s}{4}$, so $ND = CD - CN = \frac{3s}{4}$, and $CN{:}ND = \boxed{1{:}3}$.

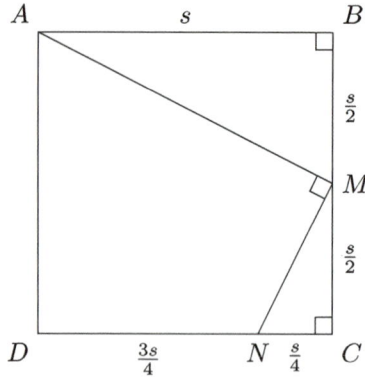

A s B

$\frac{s}{2}$

M

$\frac{s}{2}$

D $\frac{3s}{4}$ N $\frac{s}{4}$ C

Figure 3.31

45. Let $a = BC$ and $c = AB = AC$. The desired ratio is $c{:}a$.

Solution 1: Since quadrilateral $MNCB$ has a circle inscribed in it (see Fig. 3.32), its opposite sides have equal sums: $MN + BC = MB + NC$, or $\frac{3a}{2} = c$. Thus the ratio is $\boxed{3{:}2}$.

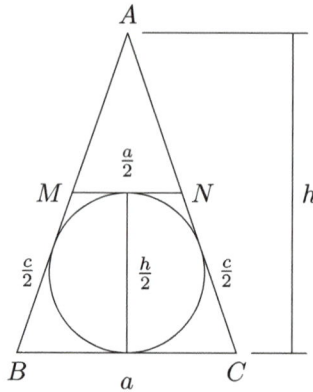

Figure 3.32

Solution 2: Let h be the length of the altitude from A to $\overline{BC}$.

$$k = rs$$

$$\frac{ha}{2} = \frac{h}{4} \cdot \left(\frac{a}{2} + c\right)$$

$$2a = \frac{a}{2} + c$$

$$\frac{3a}{2} = c.$$

Thus $c{:}a = \boxed{3{:}2}$.

46. **Solution 1:** Let R be the radius of the large circle and r be the radius of the small circle (see Fig. 3.33). If we draw the radius of the small circle tangent to the chord and the radius of the large circle to an endpoint of the chord, we have $R^2 - r^2 = 36$ by the Pythagorean theorem. The area of the region between the circles is $\pi R^2 - \pi r^2 = \boxed{36\pi}$.

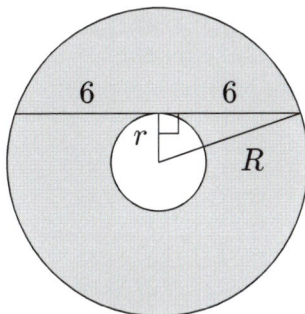

Figure 3.33

Solution 2: Since the answer is expected to be unique, we can take a special case. Assume the radius of the small circle is 0. Then the chord is a diameter of the large circle, which therefore has radius 6. The area between the circles is then just the area of the large circle, which is $\pi \cdot 6^2 = \boxed{36\pi}$.

47. Let $x = m\angle A$ (see Fig. 3.34). We repeatedly use the facts that (a) the measure of an exterior angle of a triangle equals the sum of the measures of the two remote interior angles and (b) the two base angles of an isosceles triangle are congruent to obtain the following:

$$m\angle FEA = m\angle FAE = x \qquad \text{by (b)}$$
$$m\angle DFE = m\angle FEA + m\angle FAE = 2x \qquad \text{by (a)}$$

$$m\angle FDE = m\angle DFE = 2x \qquad \text{by (b)}$$
$$m\angle CED = m\angle EAD + m\angle EDA = 3x \qquad \text{by (a)}$$
$$m\angle ECD = m\angle CED = 3x \qquad \text{by (b)}$$
$$m\angle CDB = m\angle CAB + m\angle DCA = 4x \qquad \text{by (a)}$$
$$m\angle CBD = m\angle CDB = 4x \qquad \text{by (b)}$$
$$m\angle ACB = m\angle ABC = 4x. \qquad \text{by (b)}$$

Finally, since the sum of the angles of $\triangle ABC$ must equal $180°$, we have $x + 4x + 4x = 180°$, so $x = \boxed{20°}$.

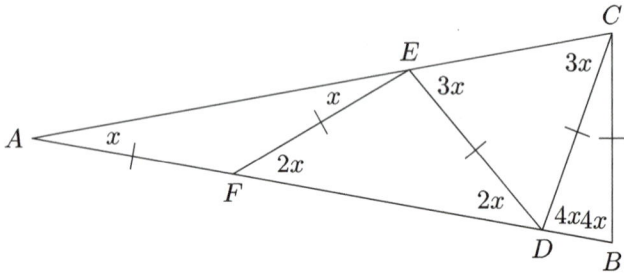

Figure 3.34

48. By alternate interior angles and angle bisectors, $\angle DQB \cong \angle QBC \cong \angle DBQ$ and $\angle EQC \cong \angle QCF \cong \angle ECQ$ (see Fig. 3.35). Thus $\triangle BDQ$ and $\triangle CEQ$ are isosceles, so $DQ = DB = 8$ and $EQ = EC = 6$. Finally, $DE = DQ - EQ = \boxed{2}$.

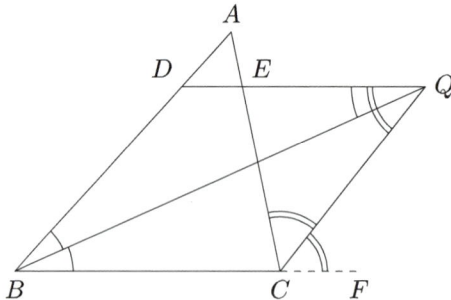

Figure 3.35

49. **Solution 1:** The length of the altitude to $\overline{BC}$ is $\sqrt{13^2 - 5^2} = 12$, so the area of $\triangle ABC$ (see Fig. 3.36) is $\frac{10 \cdot 12}{2} = 60$. That is also the sum of the areas of $\triangle APB$ and $\triangle APC$, which is $\frac{13 \cdot PX}{2} + \frac{13 \cdot PY}{2} = \frac{13}{2}(PX + PY) = 60$. Thus $PX + PY = \boxed{\frac{120}{13}}$.

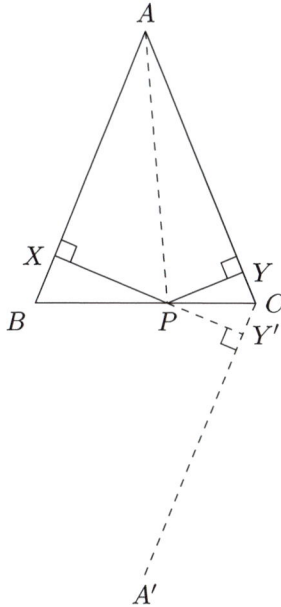

Figure 3.36

Solution 2: Reflect A and Y over $\overline{BC}$ to A' and Y'. We see $\overline{A'C} \parallel \overline{AB}$; $\overline{PX} \perp \overline{AB}$ and $\overline{PY'} \perp \overline{A'C}$, so in fact P, X, and Y' are collinear, and $PX + PY = PX + PY' = XY'$. Finally, XY' equals the length of the altitude to $\overline{AB}$, so we have $\frac{XY' \cdot 13}{2} = 60$ and $XY' = \boxed{\frac{120}{13}}$.

50. Let T be the foot of the altitude and median to $\overline{BC}$ (see Fig. 3.37). Then $BE = \frac{BC}{3}$ and $ET = BT - BE = \frac{BC}{2} - \frac{BC}{3} = \frac{BC}{6}$; let $x = ET$ and so $BE = 2x$. Then we have $h^2 + x^2 = 15^2 = 225$ and

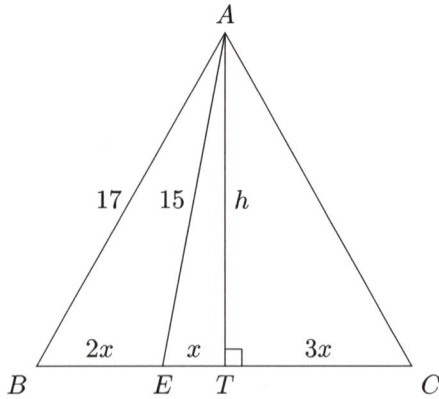

Figure 3.37

$h^2 + (3x)^2 = h^2 + 9x^2 = 17^2 = 289$. Subtracting and dividing gives $x^2 = 8$, so $x = 2\sqrt{2}$ and $BC = \boxed{12\sqrt{2}}$.

51. Note that $\overline{MN} \| \overline{AB}$; thus $\triangle DMQ \sim \triangle DAB$ and $\triangle CPN \sim \triangle CAB$, with ratio of similitude $\frac{1}{2}$ in both cases. Therefore, $MQ = \frac{AB}{2} = \frac{7}{2}$ and $MP = \frac{CD}{2} = \frac{13}{2}$, so $PQ = MP - MQ = \boxed{3}$ (see Fig. 3.38).

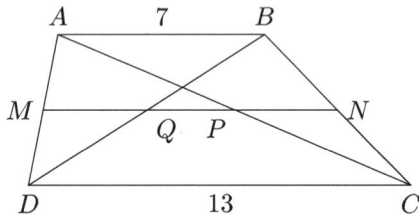

Figure 3.38

52. Extend $\overline{BA}$ through A until it intersects O_2 at D (see Fig. 3.39). Since $\triangle BAO_1$ and $\triangle DAO_2$ are both isosceles and $\angle BAO_1 \cong \angle DAO_2$, those two triangles are similar with a ratio of similitude of $O_1A:O_2A = 16:9$. Thus $AD = \frac{9}{16}AB = \frac{9}{2}$. By the power of a

point theorem, we then have $BC^2 = BA \cdot BD = 8 \cdot \left(8 + \frac{9}{2}\right) = 100$, so $BC = \boxed{10}$.

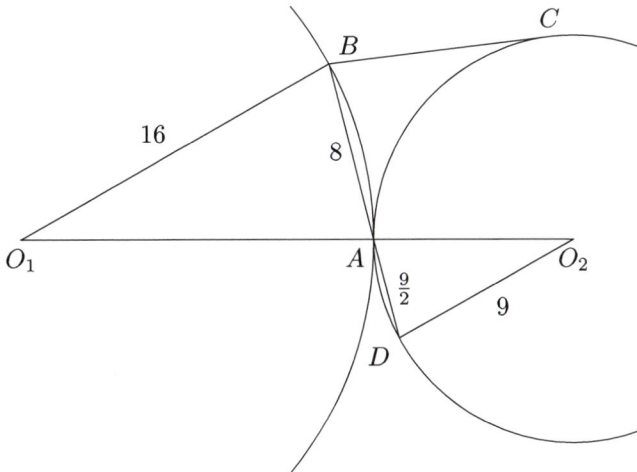

Figure 3.39

53. Since $\overline{MD}$ is part of the perpendicular bisector of $\overline{MD}$, $AD = DC$ (see Fig. 3.40); by the Pythagorean theorem, $AD = 10$, so $BC = BD + DC = 16$ and $AB = \sqrt{8^2 + 16^2} = \boxed{8\sqrt{5}}$.

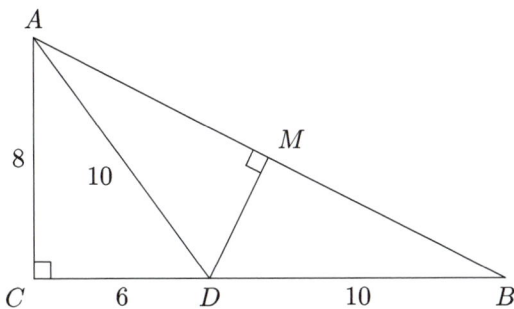

Figure 3.40

54. Since the circle is tangent to both axes, its center must lie on the bisector of the angle between them, which is the line $y = x$ (or $y = -x$, but then it would be impossible for the circle to contain

$(3, 2)$). So we can write the center as (r, r), and the distance to $(3, 2)$ equals r:

$$r = \sqrt{(r-3)^2 + (r-2)^2}$$
$$r^2 = 2r^2 - 10r + 13$$
$$0 = r^2 - 10r + 13.$$

The two possible radii are the roots of this equation, and their sum is $-\frac{-10}{1} = \boxed{10}$.

55. $\overline{FB}$, $\overline{BD}$, and $\overline{DF}$ are all diagonals of faces of the cube (see Fig. 3.41), so they are all congruent and $\triangle FBD$ is equilateral. Hence, $m\angle FBD = \boxed{60°}$.

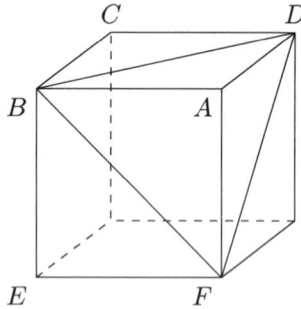

Figure 3.41

56. For a quadrilateral with an inscribed circle, the following are true: (a) $k = rs$, where k is the area of the polygon, s its semiperimeter, and r the radius of the inscribed circle, and (b) $a + c = b + d = s$, where a, b, c, and d are the lengths of the sides in order. Since the two legs of a trapezoid are opposite sides, $s = 20$, so $k = 2 \cdot 20 = \boxed{40}$.

57. Note (see Fig. 3.42) that the cube is symmetric by reflection across the plane containing Q, S, B, and D. Therefore, the reflection of E across that plane, which is the midpoint of $\overline{QR}$ (marked as F), is in the plane containing E, A, and C. Thus $ACEF$ is the trapezoid depicted on the right in Fig. 3.42. The Pythagorean theorem gives its bases as $EF = 2\sqrt{2}$ and $AC = 4\sqrt{2}$, with legs of $2\sqrt{5}$. This means its height is $3\sqrt{2}$ and its area is $\frac{2\sqrt{2}+4\sqrt{2}}{2} \cdot 3\sqrt{2} = \boxed{18}$.

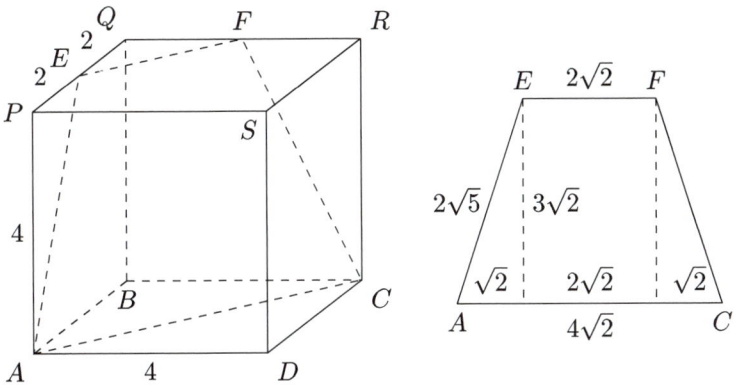

Figure 3.42

58. Let centers of the circles be A and B. The radii to the tangent points, C and D, are perpendicular to the tangent (see Fig. 3.43). Draw $\overline{BP}$ parallel to $\overline{CD}$ with P on $\overline{AD}$. Then $BPDC$ is a rectangle, so $BP = CD = 6$. Then AP, which is the difference between the radii, is $\sqrt{7^2 - 6^2} = \boxed{\sqrt{13}}$.

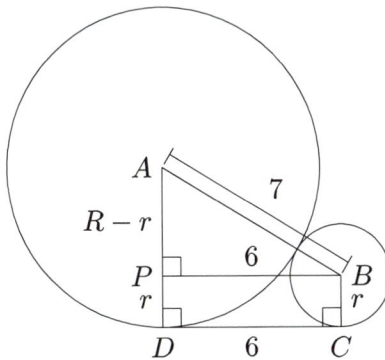

Figure 3.43

59. Let $x = m\angle PQR$ (see Fig. 3.44). Since, $PA = PQ$, $m\angle PAQ = x$ and $m\angle APQ = 180° - 2x$. By symmetry, $m\angle BPS = 180° - 2x$. so $m\angle QPS = 420 - 4x$. Finally, $180° = m\angle PQR + m\angle QPS = 420° - 3x$, so $x = \boxed{80°}$.

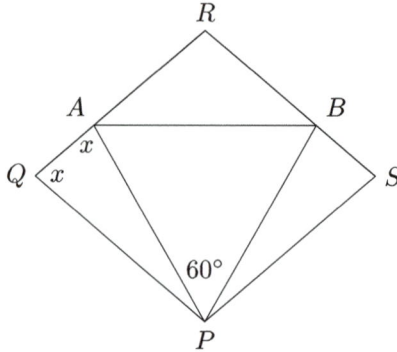

Figure 3.44

60. Simply use the Pythagorean theorem in three dimensions, since adjacent edges of a cube are perpendicular: $\sqrt{4^2 + 4^2 + 4^2} = \boxed{4\sqrt{3}}$.

61. The segment between any two opposite vertices of the cube is a diameter of the sphere, and so has length 12; by the Pythagorean theorem in three dimensions, this is $\sqrt{3}$ times the edge length of the cube, so the edge length is $\frac{12}{\sqrt{3}} = \boxed{4\sqrt{3}}$.

62. **Solution 1:** Let $x = m\angle A_1B_1C_1$ (see Fig. 3.45). $\angle A_1B_1C_1$ is an inscribed angle of $\overarc{A_1C_1}$ (on the side not including B_1), so $m\,\overarc{A_1C_1} = 2x$, and $m\,\overarc{A_1B_1C_1} = 360° - 2x$. Thus, $m\,\overarc{C_2B_1A_2} = \frac{360° - 2x}{2} = 180° - x$; $\angle A_2B_2C_2$ is an inscribed angle of that arc, so $m\angle A_2B_2C_2 = \frac{180° - x}{2} = 90° - \frac{x}{2}$. $\angle A_3B_3C_3$ is related to $\angle A_2B_2C_2$ in the same way that $\angle A_2B_2C_2$ is related to $\angle A_1B_1C_1$, so $m\angle A_3B_3C_3 = 90° - \frac{90° - \frac{x}{2}}{2} = 45° + \frac{x}{4}$. Finally, we have $x = 45° + \frac{x}{4}$, so $x = \boxed{60°}$.

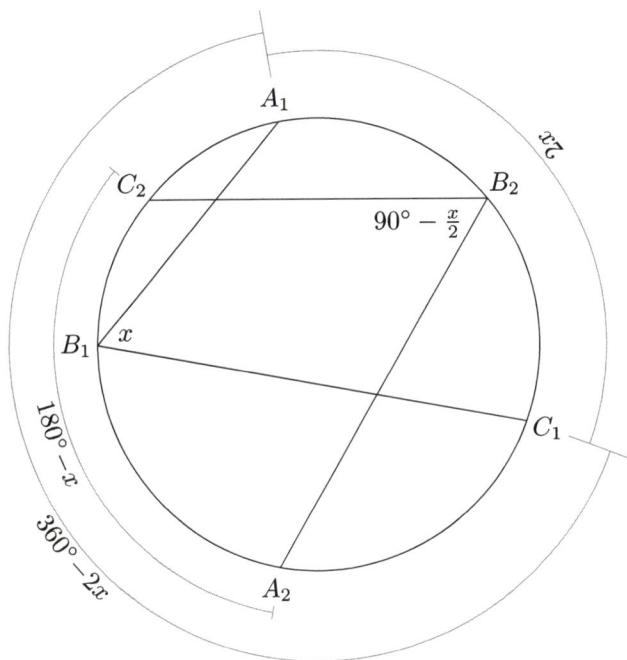

Figure 3.45

Solution 2: Since the answer is expected to be unique, we can take a special case. If $\triangle A_1 B_1 C_1$ is equilateral, so are $\triangle A_2 B_2 C_2$ and $\triangle A_3 B_3 C_3$ by symmetry, and $m\angle A_3 B_3 C_3 = m\angle A_1 B_1 C_1 = \boxed{60°}$.

63. Since $AP = PB = BA = 2$, $\triangle APB$ is equilateral, so $m\angle APB = \boxed{60°}$.

64. Let x and y represent half the lengths of the two diagonals. Then the sum of the diagonals is $2x + 2y = 14$, so $x + y = 7$. The area of the rhombus is half the product of the diagonals: $\frac{1}{2} \cdot 2x \cdot 2y = 2xy = 13$. The side length of the rhombus is $\sqrt{x^2 + y^2} = \sqrt{(x+y)^2 - 2xy} = \boxed{6}$.

Level 3

65. **Solution 1:** Use Menelaus's theorem, with $\triangle ABD$ as the triangle and $\overline{CEF}$ as the transversal (see Fig. 3.46): $1 = \frac{AC}{CD} \frac{DE}{EB} \frac{BF}{FA} =$

$2 \cdot 1 \cdot \frac{BF}{FA}$. Thus $FA = 2BF = 10$ and $AB = FA + BF = \boxed{15}$.

Solution 2: Choose point M on $\overline{EC}$ such that E bisects $\overline{MF}$. Then $FDMB$ is a parallelogram, since its diagonals bisect each other, so $DM = BF = 5$. Since $\overline{FA}\|\overline{FB}\|\overline{MD}$, $\triangle CFA \sim \triangle CMD$, and since D is the midpoint of $\overline{AC}$, the ratio of similitude is 2. Thus $AF = 2MD = 10$, and $AB = AF + FB = \boxed{15}$.

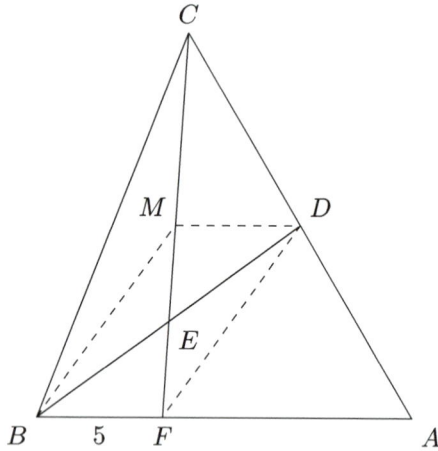

Figure 3.46

66. By use of the Pythagorean theorem on the various triangles in Fig. 3.47, we get

$$BP^2 + PX^2 = BQ^2 + QX^2,$$
$$CQ^2 + QX^2 = CR^2 + RX^2,$$
$$AR^2 + RX^2 = AP^2 + PX^2.$$

Add the three equations together and cancel common terms:

$$AP^2 + CR^2 + BQ^2 = AR^2 + CQ^2 + BP^2,$$
$$AP^2 + 16 + 4 = 25 + 9 + 1,$$
$$AP^2 = 15,$$
$$AP = \boxed{\sqrt{15}}.$$

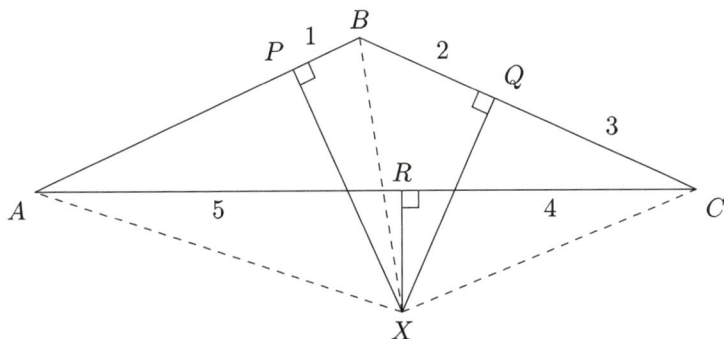

Figure 3.47

67. Let x and y be half of the lengths of the diagonals of the rhombus.

Solution 1: Note that both $\triangle ABC$ and $\triangle ABD$ have areas of xy. Each of them has two sides of length $\sqrt{x^2 + y^2}$ and a third side of length either $2x$ or $2y$ (see Fig. 3.48). Equating the areas of $\triangle ABC$ and $\triangle ABD$ and using the formula $k = \frac{abc}{4R}$ gives $xy = \frac{(x^2+y^2) \cdot 2x}{4 \cdot 2} = \frac{(x^2+y^2) \cdot 2y}{4 \cdot 6}$. The right equality gives $y = 3x$, and substituting back into the left equality gives $3x^2 = \frac{5x^3}{2}$, so $x = \frac{6}{5}$, $y = \frac{18}{5}$, and the area of the rhombus is $\boxed{\frac{216}{25}}$.

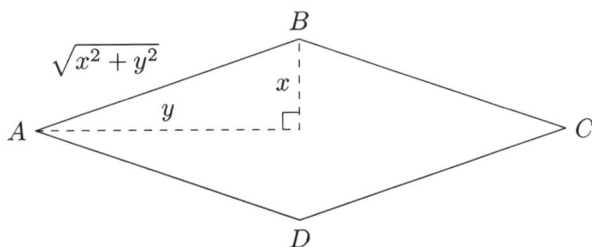

Figure 3.48

Solution 2: Consider finding the circumcircle of an isosceles triangle $\triangle ABC$ with base $2a$ and height h, as shown in Fig. 3.49. The circumcenter O is the intersection of the sides' perpendicular bisectors, and the circumradius is AO. Let N be the midpoint of $\overline{BC}$; then $\overline{AN}$ is one of the bisectors. Let M be the midpoint of

$\overline{AB}$, so that $\overline{OM}$ is part of the perpendicular bisector of $\overline{AB}$ and therefore perpendicular to $\overline{AB}$. Then we have $\triangle ANB \sim \triangle AMO$, so $\frac{AO}{AM} = \frac{AB}{AN}$, or $AO = \frac{AM \cdot AB}{AN} = \frac{a^2+h^2}{2h}$. Then, for the two isosceles triangles in the problem, we have $\frac{x^2+y^2}{2x} = 6$ and $\frac{x^2+y^2}{2y} = 2$, so $y = 3x$. Substituting back in gives $6 = \frac{x^2+9x^2}{2x} = 5x$, so $x = \frac{6}{5}$ and $y = \frac{18}{5}$. The area of the rhombus is $2xy = \boxed{\frac{216}{25}}$.

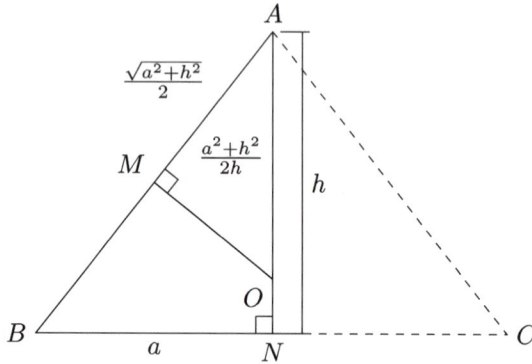

Figure 3.49

68. **Solution 1:** Since all three angle bisectors of any triangle coincide at one point, $\overline{MQ}$ bisects $\angle M$ (see Fig. 3.50). Thus $\widehat{PQ} \cong \widehat{QN}$, and so $PQ = QN$. Let $\alpha = m\angle K$, $\beta = m\angle L$, $\gamma = m\angle M$, $\theta = m\angle MPQ$, and $\phi = m\angle MNQ$. Since θ and ϕ are the inscribed angles of two arcs which add up to $360°$, $\theta + \phi = 180°$. Consider $\triangle KLM$, $\triangle MNK$, and $\triangle MPL$; since the interior angles of any triangle add up to $180°$, we have

$$\theta + \phi = 180° \tag{1}$$

$$\alpha + \beta + \gamma = 180° \tag{2}$$

$$\gamma + \frac{\alpha}{2} + \phi = 180° \tag{3}$$

$$\gamma + \frac{\beta}{2} + \theta = 180° \tag{4}$$

$$2\gamma + \frac{\alpha}{2} + \frac{\beta}{2} = 180° \qquad (5) = (3) + (4) - (1)$$

$$4\gamma + \alpha + \beta = 360° \qquad\qquad (6) = 2 \cdot (5)$$
$$3\gamma = 180° \qquad\qquad (7) = (6) - (2)$$
$$\gamma = 60°.$$

Thus $m\angle PQN = 120°$, so $\triangle PQN$ is as depicted to the right, so $PQ = \boxed{\dfrac{2\sqrt{3}}{3}}$.

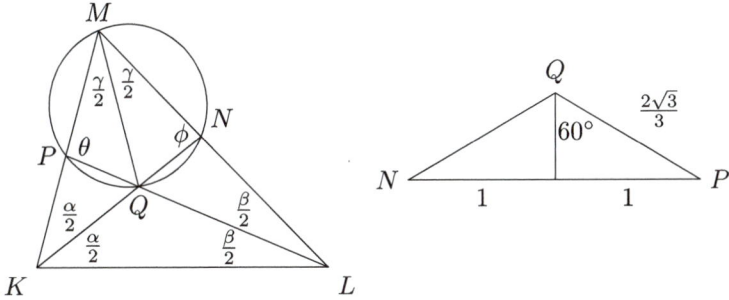

Figure 3.50

Solution 2: Since the answer is expected to be unique, we can take a special case. Assume $\triangle KLM$ is equilateral (see Fig. 3.51). Then $m\angle QPM = m\angle QNM = 90°$, so the circle which has $\overline{MQ}$ as a diameter contains both P and N, so this satisfies the conditions given in the problem. N and P are the midpoints of two sides of $\triangle KLM$, so the side length of $\triangle KLM$ is $2PN = 4$. PQ is the distance from the midpoint of a side to the centroid, which is one third the length of a median, so $PQ = \boxed{\dfrac{2\sqrt{3}}{3}}$.

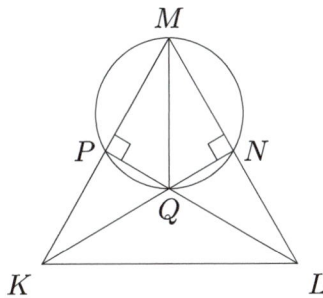

Figure 3.51

69. The conditions of the problem mean that $\triangle ABC$ and $\triangle ADC$ have equal areas, and that B and D are on opposite sides of $\overleftrightarrow{AC}$. If two vertices and the area of a triangle are fixed, the possible positions of the other vertex make up two lines parallel to the line containing the two given vertices, and at equal distances from it. In this case, the fixed vertices A and C lie on the line $x = y$, and B lies on the parallel line $x = y - 31$. D must lie on the reflection of this line over $x = y$, which is the graph of the functional inverse, so the reflection is described by $y = x - 31$, or $x - y = \boxed{31}$.

70. Let $x = AE = ED$ and $y = EC$ (see Fig. 3.52). Use the law of cosines in $\triangle EBD$ and then in $\triangle AEC$:

$$x^2 = 5^2 + (15 - x)^2 - 2 \cdot 5 \cdot (15 - x) \cdot \cos 60°$$
$$= 25 + 225 - 30x + x^2 - 75 + 5x$$
$$25x = 175$$
$$x = 7;$$
$$y^2 = x^2 + 15^2 - 2 \cdot x \cdot 15 \cdot \cos 60°$$
$$= x^2 - 15x + 225$$
$$= 169$$
$$y = \boxed{13}.$$

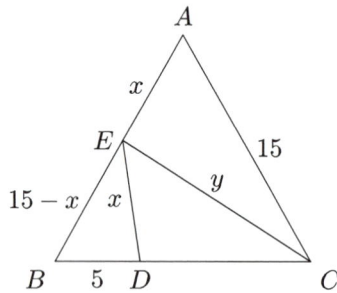

Figure 3.52

71. By the Pythagorean theorem, $DB = 26$. Assume that B and D are the vertices brought together by the fold. The crease (denoted $\overline{XY}$ in Fig. 3.53) must be the perpendicular bisector of $\overline{BD}$: after folding, B and D are the same point, so any point on the crease is

the same distance from them, but doing the fold does not change the distance between any two points if the line segment between them does not cross the crease, so any point on the crease is equidistant from B and D to begin with. Thus $BM = 13$ and $\triangle ABD \sim \triangle MBX$ with ratio of similitude $AB{:}MB = 24{:}13$. Thus $XM = \frac{13}{24}$, $AD = \frac{65}{12}$. By symmetry, $YM = XM$, so $XY = 2XM = \boxed{\frac{65}{6}}$.

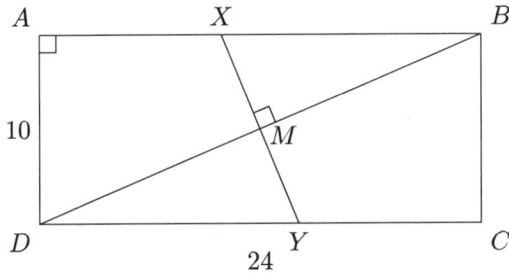

Figure 3.53

72. The polygon has threefold rotational symmetry about the center of the circle, so $\triangle ACE$ is an equilateral triangle inscribed in the circle (see Fig. 3.54), $m\,\widehat{AEC} = 240°$, and $m\angle ABC = 120°$. Using the law of cosines in $\triangle ABC$ gives $AC^2 = 2^2 + 10^2 - 2\cdot 2\cdot 10\cdot \cos 120° = 124$. The area of $\triangle ACE$ is therefore $\frac{124\sqrt{3}}{4} = \boxed{31\sqrt{3}}$.

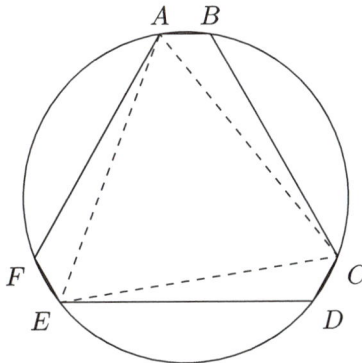

Figure 3.54

Chapter 4

Trigonometry

Questions

Level 1

1. Compute the numerical value of $\sin^2 \frac{\pi}{10} + \sin^2 \frac{2\pi}{5}$.
2. The roots of the equation $x^2 - px + q = 0$ are $\tan\theta$ and $\cot\theta$. Compute the numerical value of q.
3. Compute the degree measure of the largest angle of a triangle whose sides have lengths 3, 5, and 7.
4. In $\triangle ABC$, $m\angle C = 90°$. Compute the value of $\cot A \cdot \cot B$.
5. If $\sin x = \frac{1}{4}$, compute all possible values of $\sin 2x$.
6. Square $ABMN$ is constructed on the hypotenuse of right triangle $\triangle ABC$. If $AC = 1$ and $BC = 20$, compute MC.
7. For some x, the legs of a right triangle are $\sin x$ and $\cos x$. Compute the circumradius of this triangle.
8. Compute $\cot\left(\arcsin \frac{1}{\pi} + \arccos \frac{1}{\pi}\right)$.
9. Compute $\cos\left(\arcsin \frac{4}{5} + \arctan \frac{5}{12}\right)$.
10. Express $\sin 345°$ in simplest radical form.
11. Find the maximum value of $5\sin 4x \cos 4x$.
12. Compute $\sin \frac{\pi}{7} + \sin \frac{4\pi}{7} + \sin \frac{7\pi}{7} + \sin \frac{10\pi}{7} + \sin \frac{13\pi}{7}$.

Level 2

13. If $\sin x + \cos x = \frac{1}{2}$, compute $\sin 2x$.
14. Compute all x such that $x \in [0°, 360°]$ and $\frac{1 - \cos 2x}{\sin 2x} = 1$.

15. Compute all x such that $x \in [0°, 360°]$ and $1 + 2\sin x + 2\cos x + \sin 2x + \cos 2x = 0$.

16. Compute all x such that $x \in [0°, 360°]$ and $\sin 2x < \sin x$.

17. Compute x such that $x \in [0°, 90°]$ and $\cos x = \cos 40° + \cos 80°$.

18. Compute $(\sin 165° - \sin 75°)^8$.

19. If $\tan^2(180° - x) + \sec(180° + x) = 11$, compute all possible values of $\cos x$.

20. In $\triangle ABC$, $\sin A : \sin B : \sin C = 5 : 7 : 9$. Compute $\cos(A + B)$.

21. Compute x such that $x \in [0°, 90°]$ and $\cos^4 x + \sin^4 x = \frac{3}{4}$.

22. Compute $\frac{\sin 75° + \cos 75°}{\sin 75° - \cos 75°}$.

23. Express in simplest radical form: $\tan 18° + \tan 42° + \sqrt{3}\tan 18° \tan 42°$.

24. Find $\tan x$ if $\sin x + \cos x = \frac{1}{5}$ and $x \in [90°, 180°]$.

25. If $\sin 2x = \frac{1}{4}$, compute $\sin^4 x + \cos^4 x$.

26. Compute all x such that $x \in [0°, 360°]$ and $\sin 2x + \cos 2x$ is maximized.

27. If $\sin x = \frac{1}{3}$, compute $\sin^4 x - \cos^4 x$.

28. A convex equiangular octagon has sides with lengths that alternate between 1 and 2. Find the ordered pair of integers (a, b) such that the area of the circle circumscribed around this octagon has area $\frac{\pi}{2}(a + b\sqrt{2})$.

29. Compute all x such that $x \in [0°, 90°]$ and $\sin 2x \geq \cos x$.

30. Compute $\tan\left(4\arctan\frac{1}{3}\right)$.

31. Compute the distance between the origin and any intersection point of the polar graph of $r^2 = 8\sin 2\theta$ and the Cartesian graph of $y = \frac{1}{x}$.

32. Express $\sin 285°$ in simplest radical form.

33. Compute $\frac{1}{\sin 10°} - \frac{\sqrt{3}}{\cos 10°}$.

34. Compute the area of a triangle whose sides have lengths $\sqrt{5}$, $\sqrt{6}$, and $\sqrt{7}$.

35. Compute the smallest positive x such that $\tan x + \sec 2x = 1$.

36. x and y are angles such that $\cos x + \cos y = \frac{1}{2}$, $\sin x + \sin y = \frac{1}{4}$, and $\sin 2x + \sin 2y = -\frac{27}{20}$. Compute $\sin(x + y)$.

37. Compute all x such that $x \in [0°, 360°]$ and $\sin x + \cos x = 1 + \sin 2x$.

38. In $\triangle ABC$, $AB = 20$, $BC = 13$, and $AC = 21$. If $\cos A + \cos B + \cos C = \frac{p}{q}$, where p and q are positive and relatively prime, compute $p + q$.

39. The angles of a triangle with sides 3, 7, and x form an arithmetic progression. Compute all possible values of x.

Level 3

40. Compute $\cos 20° \cdot \cos 40° \cdot \cos 60° \cdot \cos 80°$.
41. Compute $\cos 10° \cdot (\cos 20° - 2\sin^2 10°)$.
42. In $\triangle ABC$, $AB = 3$, $BC = 8$, and $AC = 7$. Equilateral triangles $\triangle ABP$, $\triangle BCQ$, and $\triangle CAR$ are exterior to $\triangle ABC$. Compute $PC + QA + RB$.
43. A triangle has area 12 and two medians of length 5 and 6. Find all possible lengths of the third median.
44. Find the exact value of $\cos 72° - \cos 36°$.
45. In quadrilateral $ABCD$, $CD = 1$, $BC = 2$, $m\angle C = 120°$ and $m\angle B = m\angle D = 90°$. Find AB.
46. If $\sin^6 x + \cos^6 x = \frac{2}{3}$ and $x \in [0°, 90°]$, compute $\sin 2x$.
47. A regular n-gon is inscribed in a circle of radius r. Find all n such that the area of the n-gon is an integral multiple of r^2.
48. If a, b, and c are the sides of $\triangle ABC$ (opposite vertices A, B, and C respectively) such that a^2, b^2, and c^2 form an arithmetic progression and $\cot A + \cot C = 3$, compute $\cot B$.

Answers

Level 1

1. $\frac{\pi}{10} + \frac{2\pi}{5} = \frac{\pi}{2}$, so $\sin\frac{2\pi}{5} = \cos\frac{\pi}{10}$ and we have $\sin^2\frac{\pi}{10} + \cos^2\frac{\pi}{10} = \boxed{1}$.
2. Since q is the product of the roots, $q = \tan\theta\cot\theta = \boxed{1}$.
3. The largest angle (call it x) is opposite the largest side. The law of cosines gives $7^2 = 3^2 + 5^2 - 2 \cdot 3 \cdot 5 \cdot \cos x$ or $15 = -30\cos x$, so $\cos x = -\frac{1}{2}$; since x is an angle of a triangle, $x \in [0°, 180°]$, so $x = \boxed{120°}$.
4. $\cot A = \frac{b}{a}$ and $\cot B = \frac{a}{b}$, so $\cot A \cdot \cot B = \boxed{1}$.
5. $\cos x = \pm\sqrt{1 - \sin^2 x} = \pm\frac{\sqrt{15}}{4}$ and $\sin 2x = 2\sin x \cos x = \boxed{\pm\frac{\sqrt{15}}{8}}$.
6. **Solution 1:** The law of cosines gives

$$MC^2 = BC^2 + BM^2 - 2 \cdot BC \cdot BM \cdot \cos(m\angle MBC)$$
$$= BC^2 + AC^2 + BC^2 - 2 \cdot BC \cdot BM \cdot \cos(m\angle ABC + 90°)$$

$$= 2BC^2 + AC^2 + 2 \cdot BC \cdot BM \cdot \sin(m\angle ABC)$$

$$= 2BC^2 + AC^2 + 2 \cdot BC \cdot BM \cdot \frac{AC}{BM}$$

$$= 2BC^2 + AC^2 + 2 \cdot BC \cdot AC$$

$$= 841$$

$$MC = \boxed{29}.$$

Solution 2: Consider rotating A, B, and C by $90°$ counterclockwise around the center of $ABMN$ (see Fig. 4.1). A and B map to B and M and the image of $\overline{AC}$ is parallel to $\overline{BC}$, but since A maps to B, the image is actually an extension of $\overline{BC}$. Let P be the image of C; then $\overline{BP} \perp \overline{PM}$, $BP = AC = 1$, and $PM = CB = 20$. Finally, $CM = \sqrt{CP^2 + PM^2} = \sqrt{20^2 + 21^2} = \boxed{29}$.

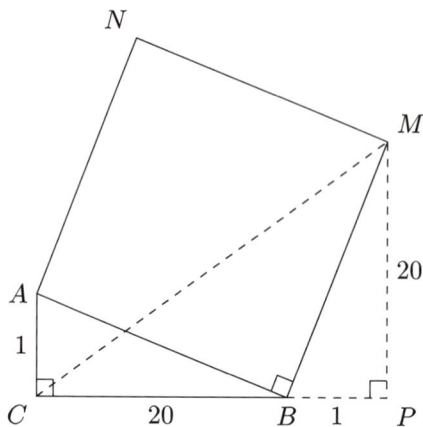

(Diagram not to scale.)

Figure 4.1

7. The hypotenuse is $\sqrt{\sin^2 x + \cos^2 x} = 1$. Since the hypotenuse of a right triangle is a diameter of its circumcircle, the radius is $\boxed{\frac{1}{2}}$.

8. Let $x = \arcsin \frac{1}{\pi}$ and $y = \arccos \frac{1}{\pi}$. x and y are acute and $\sin x = \cos y$, so x and y are complementary. Thus $\cot(x + y) = \cot 90° = \boxed{0}$.

9. Let $x = \arcsin \frac{4}{5}$ and $y = \arctan \frac{5}{12}$. x and y are acute, and $\cos x = \frac{3}{5}$, $\sin y = \frac{5}{13}$, and $\cos y = \frac{12}{13}$. Finally, $\cos(x + y) = \cos x \cos y - \sin x \sin y = \frac{3}{5} \frac{12}{13} - \frac{4}{5} \frac{5}{13} = \boxed{\frac{16}{65}}$.

10. $\sin 345° = \sin(300° + 45°) = \sin 300° \cos 45° + \cos 300° \sin 45° = -\frac{\sqrt{3}}{2} \frac{\sqrt{2}}{2} + \frac{1}{2} \frac{\sqrt{2}}{2} = \boxed{\frac{\sqrt{2} - \sqrt{6}}{4}}$.

11. $5 \sin 4x \cos 4x = \frac{5}{2} \sin 8x$, which has a maximum value of $\boxed{\frac{5}{2}}$.

12. Since $\frac{13\pi}{7} + \frac{\pi}{7} = \frac{10\pi}{7} + \frac{4\pi}{7} = 2\pi$, $\sin \frac{13\pi}{7} = -\sin \frac{\pi}{7}$ and $\sin \frac{10\pi}{7} = -\sin \frac{4\pi}{7}$. Thus all the terms cancel each other except for $\sin \frac{7\pi}{7} = \boxed{0}$.

Level 2

13. $\frac{1}{4} = (\sin x + \cos x)^2 = \sin^2 + 2 \sin x \cos x + \cos^2 = 1 + \sin 2x$, so $\sin 2x = \boxed{-\frac{3}{4}}$.

14. We have $1 = \frac{1 - (1 - 2\sin^2 x)}{2 \sin x \cos x} = \frac{\sin x}{\cos x} = \tan x$, so $x = \boxed{45° \text{ or } 225°}$.

15.

$$0 = 1 + 2 \sin x + 2 \cos x + 2 \sin x \cos x + 2 \cos^2 x - 1$$
$$= 2 \sin x + 2 \cos x + 2 \sin x \cos x + 2 \cos^2 x$$
$$= 2(\sin x + \cos x)(1 + \cos x).$$

Thus either $\sin x + \cos x = 0$ (which implies $\tan x = -1$), or $1 + \cos x = 0$, which gives the solutions $x = \boxed{135°, 180°, \text{ or } 315°}$.

16. $\sin 2x - \sin x < 0$ yields $2 \sin x \cos x - \sin x < 0$, or $(2 \cos x - 1)(\sin x) < 0$. This can be true if either (a) $\sin x < 0$ and $2 \cos x - 1 > 0$ or (b) $\sin x > 0$ and $2 \cos x - 1 < 0$. In the first case, $\sin x < 0$ implies $x \in (180°, 360°)$ and $2 \cos x - 1 > 0$ implies $x \in (0°, 60°)$ or $x \in (300°, 360°)$. In the second case, $\sin x > 0$ implies $x \in (0°, 180°)$ and $2 \cos x - 1 < 0$ implies $x \in (60°, 300°)$. Putting all of these together gives $\boxed{x \in (60°, 180°) \text{ or } x \in (300°, 360°)}$.

17.

$$\cos 40° + \cos 80° = \cos(60° - 20°) + \cos(60° + 20°)$$
$$= (\cos 60° \cos 20° + \cancel{\sin 60° \sin 20°}) +$$
$$+ (\cos 60° \cos 20° - \cancel{\sin 60° \sin 20°})$$
$$= 2\cancel{\cos 60} \cos 20$$
$$= \cos 20$$

so $x = \boxed{20°}$.

18. As in the previous problem, we have $\sin 165° - \sin 75° = \sin(120° + 45°) - \sin(120° - 45°) = 2\cos 120° \sin 45° = -\frac{\sqrt{2}}{2}$, and $\left(-\frac{\sqrt{2}}{2}\right)^8 = \boxed{\frac{1}{16}}$.

19. We have $\tan(180° - x) = -\tan x$ and $\sec(180° + x) = -\sec x$. Thus,

$$11 = \tan^2 x - \sec x$$
$$= (\sec^2 x - 1) - \sec x$$
$$0 = \sec^2 - \sec x - 12$$
$$= (\sec x - 4)(\sec x + 3).$$

Thus $\sec x = -3$ or 4, so $\cos x = \frac{1}{\sec x} = \boxed{-\frac{1}{3} \text{ or } \frac{1}{4}}$.

20. By the law of sines, the ratio of the sides of the triangle is $a : b : c = 5 : 7 : 9$. Since any two triangles with that ratio are similar, this determines the angles, and we can take the sides to be 5, 7, and 9. Then, by the law of cosines, we have $9^2 = 5^2 + 7^2 - 2 \cdot 5 \cdot 7 \cdot \cos C$, so $\cos C = -\frac{1}{10}$. Since $A + B = 180° - C$, $\cos(A + B) = -\cos C = \boxed{\frac{1}{10}}$.

21.

$$\frac{3}{4} = \cos^4 x + \sin^4 x = \left(\sin^2 x + \cos^2 x\right)^2 - 2\sin^2 x \cos^2 x$$
$$= 1 - 2\sin^2 \cos^2 x$$
$$= 1 - \frac{\sin^2 2x}{2}$$
$$\sin 2x = \pm\frac{\sqrt{2}}{2}$$

Since $2x \in [0°, 180°]$, $\sin 2x \geq 0$, so $\sin 2x = \frac{\sqrt{2}}{2}$ and $2x = 45°$ or $135°$. Thus $x = \boxed{22.5° \text{ or } 67.5°}$.

22. **Solution 1:** Square the quantity:

$$\left(\frac{\sin 75° + \cos 75°}{\sin 75° - \cos 75°}\right)^2 = \frac{\sin^2 75° + \cos^2 75° + 2\sin 75° \cos 75°}{\sin^2 75° + \cos^2 75° - 2\sin 75° \cos 75°}$$

$$= \frac{1 + \sin 150°}{1 - \sin 150°}$$

$$= \frac{3/2}{1/2} = 3.$$

Thus the original value is $\boxed{\sqrt{3}}$.

Solution 2: If you happen to recall that $\sin 75° = \frac{\sqrt{6}+\sqrt{2}}{4}$ and $\cos 75° = \frac{\sqrt{6}-\sqrt{2}}{4}$, then

$$\frac{\sin 75° + \cos 75°}{\sin 75° - \cos 75°} = \frac{\sqrt{6} + \sqrt{2} + \sqrt{6} - \sqrt{2}}{\sqrt{6} + \sqrt{2} - \sqrt{6} + \sqrt{2}} = \frac{2\sqrt{6}}{2\sqrt{2}} = \boxed{\sqrt{3}}.$$

23.

$$\sqrt{3} = \tan 60° = \tan(18° + 42°)$$

$$= \frac{\tan 18° + \tan 42°}{1 - \tan 18° \tan 42°}$$

$$\sqrt{3} - \sqrt{3}\tan 18° \tan 42° = \tan 18° + \tan 42°$$

$$\boxed{\sqrt{3}} = \tan 18° + \tan 42° + \sqrt{3}\tan 18° \tan 42°.$$

24. **Solution 1:**

$$\sin x + \cos x = \frac{1}{5}$$

multiplying both sides of the above equation by 5 and then dividing by $\cos x$, we get

$$5\tan x + 5 = \sec x$$

$$25\tan^2 x + 50\tan x + 25 = \sec^2 x$$

$$= \tan^2 x + 1$$

$$24\tan^2 x + 50\tan x + 24 = 0$$

$$(3\tan x + 4)(4\tan x + 3) = 0.$$

So $\tan x = -\frac{4}{3}$ or $\tan x = -\frac{3}{4}$. But if $\tan x = -\frac{3}{4}$, then, since x is in the second quadrant, $\sin x + \cos x = \frac{3}{5} - \frac{4}{5} \neq \frac{1}{5}$. So we must have

$$\tan x = \boxed{-\frac{4}{3}}.$$

Solution 2: Let $d = \frac{1}{5}$. The point $(\cos x, \sin x)$ is the intersection, in the second quadrant, of the line $x + y = d$ and the unit circle centered at the origin; call this point P. Let O be the origin and M be the point $\left(\frac{d}{2}, \frac{d}{2}\right)$; then M is on the line and $\overline{OM} \perp \overline{MP}$ (see Fig. 4.2). We have $OM = \frac{d}{\sqrt{2}}$ and $OP = 1$, so $MP = \sqrt{1 - \frac{d^2}{2}}$. The vector from M to P is $\left(-\frac{1}{\sqrt{2}}, \frac{1}{\sqrt{2}}\right) \cdot MP = \left(-\frac{\sqrt{2-d^2}}{2}, \frac{\sqrt{2-d^2}}{2}\right)$. Thus $P = \left(\frac{d}{2} - \frac{\sqrt{2-d^2}}{2}, \frac{d}{2} + \frac{\sqrt{2-d^2}}{2}\right) = \left(-\frac{3}{5}, \frac{4}{5}\right)$, and $\tan x = \frac{4/5}{-3/5} = \boxed{-\frac{4}{3}}.$

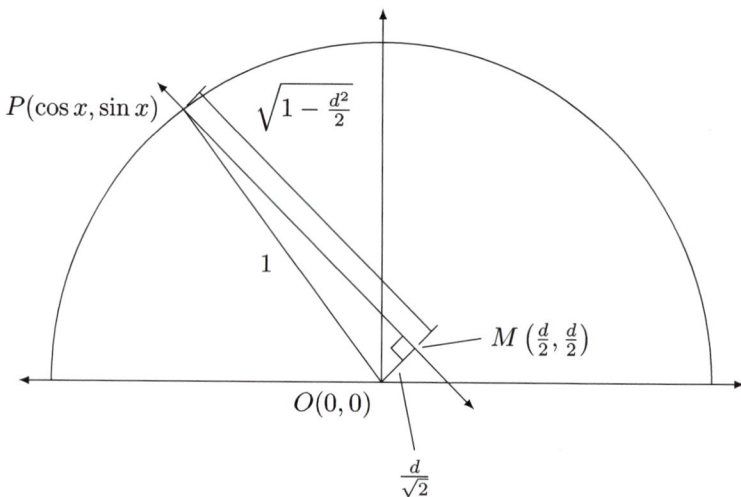

$P(\cos x, \sin x)$ $\sqrt{1 - \frac{d^2}{2}}$

1

$M\left(\frac{d}{2}, \frac{d}{2}\right)$

$O(0,0)$

$\frac{d}{\sqrt{2}}$

Figure 4.2

25.

$$1 = (\sin^2 x + \cos^2 x)^2$$
$$= \sin^4 x + \cos^4 x + 2\sin^2 x \cos^2 x$$
$$= \sin^4 x + \cos^4 x + \frac{\sin^2 2x}{2}$$

$$= \sin^4 x + \cos^4 x + \frac{1}{32}$$

$$\boxed{\frac{31}{32}} = \sin^4 x + \cos^4 x.$$

26. $\sin 2x + \cos 2x = \sqrt{2}\sin(2x + 45°)$. As x goes from $0°$ to $360°$, $2x + 45°$ goes from $45°$ to $765°$, and $\sin(2x + 45°)$ is maximized in this range when $2x + 45° = 90°$ or $450°$, which means $x = \boxed{22.5° \text{ or } 202.5°}$.

27. $\sin^4 x - \cos^4 x = \left(\cancel{\sin^2 x + \cos^2 x}\right)\left(\sin^2 x - \cos^2 x\right) = 2\sin^2 x - 1 = \boxed{-\frac{7}{9}}$.

28. The polygon has fourfold rotational symmetry about the center of the circle, so $ACEG$ is a square, $m\,\widehat{AGEC} = 270°$, and $m\angle ABC = 135°$ (see Fig. 4.3). Using the law of cosines in $\triangle ABC$ gives $AC^2 = 1^2 + 2^2 - 2 \cdot 1 \cdot 2 \cdot \cos 135° = 5 + 2\sqrt{2}$. $\triangle AOC$ is a $45 - 45 - 90$ right triangle, so $AO^2 = \frac{AC^2}{2}$. $\overline{AO}$ is a radius of the circle, so the area of the circle is $\pi AO^2 = \frac{\pi}{2}AC^2 = \frac{\pi}{2}\left(5 + 2\sqrt{2}\right)$, so the answer is $\boxed{(5, 2)}$.

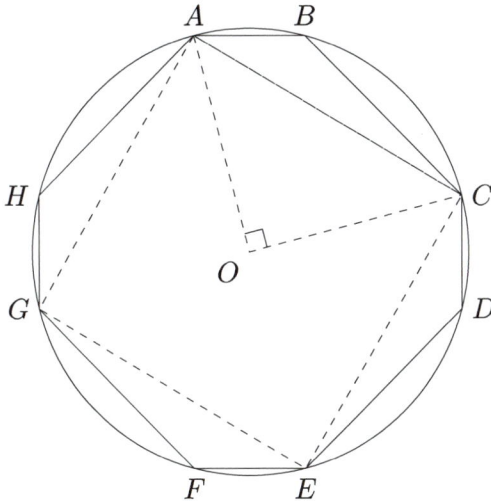

Figure 4.3

29. $\sin 2x = 2 \sin x \cos x$, so the inequality becomes $2 \sin x \cos x - \cos x \geq 0$, or $\cos x(2 \sin x - 1) \geq 0$. Since x is in the first quadrant, $\cos x \geq 0$, so we need $2 \sin x - 1 \geq 0$, or $\sin x \geq \frac{1}{2}$, which is true when $x \in [30°, 150°]$. So, restricting to the first quadrant, the valid values are $\boxed{x \in [30°, 90°]}$.

30. Let $x = \arctan \frac{1}{3}$, so that $\tan x = \frac{1}{3}$ and we want $\tan 4x$:

$$\tan 2x = \frac{2 \tan x}{1 - \tan^2 x} = \frac{3}{4} \quad \text{and} \quad \tan 4x = \frac{2 \tan 2x}{1 - \tan^2 2x} = \boxed{\frac{24}{7}}.$$

31. We have $x = r \cos \theta$, $y = r \sin \theta$, and $r^2 = x^2 + y^2$. For any intersection point,

$$r^2 = 8 \sin 2\theta$$
$$= 16 \sin \theta \cos \theta$$
$$r^4 = 16(r \cos \theta)(r \sin \theta)$$
$$(x^2 + y^2)^2 = 16xy$$
$$= 16.$$

Thus the distance from the origin to any intersection point is $\sqrt{x^2 + y^2} = \sqrt[4]{16} = \boxed{2}$.

32. $\sin 285° = \sin(240° + 45°) = \sin 240° \cos 45° + \cos 240° \sin 45° = -\frac{\sqrt{3}}{2} \cdot \frac{\sqrt{2}}{2} - \frac{1}{2} \cdot \frac{\sqrt{2}}{2} = \boxed{\frac{-\sqrt{6} - \sqrt{2}}{4}}$.

33. Call the expression M. Then $M = \frac{\cos 10° - \sqrt{3} \sin 10°}{\sin 10° \cos 10°}$ and dividing both sides by 4, $\frac{M}{4} = \frac{\frac{1}{2} \cos 10° - \frac{\sqrt{3}}{2} \sin 10°}{2 \sin 10° \cos 10°} = \frac{\sin 30° \cos 10° - \cos 30° \sin 10°}{\sin 20°} = \frac{\sin(30° - 10°)}{\sin 20°} = 1$. So $M = \boxed{4}$.

34. **Solution 1:** Let x be the angle between the two shortest sides. By the law of cosines, $\cos x = \frac{5 + 6 - 7}{2 \cdot \sqrt{5} \cdot \sqrt{6}} = \frac{2}{\sqrt{30}}$, and so $\sin x = \sqrt{1 - \cos^2 x} = \sqrt{\frac{13}{15}}$. Then the area of the triangle is $\frac{1}{2} \cdot \sqrt{5} \cdot \sqrt{6} \cdot \sin x = \boxed{\frac{\sqrt{26}}{2}}$.

Solution 2: Symbolically expand Heron's formula, keeping in mind that $(a+b)(a-b) = a^2 - b^2$:

$$k = \frac{1}{4}\sqrt{(a+b+c)(-a+b+c)(a-b+c)(a+b-c)}$$

$$= \frac{1}{4}\sqrt{(-a^2+b^2+c^2+2bc)(a^2-b^2-c^2+2bc)}$$

$$= \frac{1}{4}\sqrt{4b^2c^2 - a^4 - b^4 - c^4 + 2a^2b^2 + 2a^2c^2 - 2b^2c^2}$$

$$= \frac{1}{4}\sqrt{2(a^2b^2 + a^2c^2 + b^2c^2) - a^4 - b^4 - c^4}.$$

Here, a^2, b^2, and c^2 are 5, 6, and 7, so this is $\frac{1}{4}\sqrt{104} = \boxed{\frac{\sqrt{26}}{2}}$.

Solution 3: Use Heron's formula, keeping in mind that $(a+b)(a-b) = a^2 - b^2$:

$$k = \frac{1}{4}\sqrt{(\sqrt{5}+\sqrt{6}+\sqrt{7})(-\sqrt{5}+\sqrt{6}+\sqrt{7})(\sqrt{5}-\sqrt{6}+\sqrt{7})}$$

$$\times \sqrt{(\sqrt{5}+\sqrt{6}-\sqrt{7})}$$

$$= \frac{1}{4}\sqrt{(6+2\sqrt{42}+7-5)(5-(6+7-2\sqrt{42}))}$$

$$= \frac{1}{4}\sqrt{(8+2\sqrt{42})(-8+2\sqrt{42})}$$

$$= \frac{1}{4}\sqrt{104}$$

$$= \boxed{\frac{\sqrt{26}}{2}}.$$

35.

$$\frac{\sin x}{\cos x} + \frac{1}{\cos 2x} = 1$$

$$\sin x \cos 2x + \cos x = \cos x \cos 2x$$

$$\sin x \cos 2x = \cos x \left(\cos 2x - 1\right)$$

$$= -2\cos x \sin^2 x$$

$$\sin x \left(\cos 2x + 2\cos x \sin x\right) = 0$$

$$\sin x \left(\cos 2x + \sin 2x\right) = 0$$

The smallest positive x for which $\sin x = 0$ is $180°$, and the smallest positive $2x$ for which $\cos 2x + \sin 2x = 0$, i.e., $\tan 2x = -1$, is $135°$, or $x = \boxed{67.5°}$.

36.

$$\frac{1}{8} = (\cos x + \cos y)(\sin x + \sin y)$$

$$= \cos x \sin x + \cos x \sin y + \cos y \sin x + \cos y \sin y$$

$$= \frac{1}{2}(\sin 2x + \sin 2y) + \sin(x + y)$$

$$= -\frac{27}{40} + \sin(x + y)$$

$$\boxed{\frac{4}{5}} = \sin(x + y).$$

37. First note that $0 \leq 1 + \sin 2x = \sin x + \cos x$. Then, $(1 + \sin 2x)^2 = (\sin x + \cos x)^2 = \sin^2 x + \cos^2 x + 2 \sin x \cos x = 1 + \sin 2x$, so $1 + \sin 2x = 1$ or 0, so $\sin 2x = -1$ or 0, which is satisfied by $x = 0°, 90°, 135°, 180°, 270°$, or $315°$. The additional constraint that $\sin x + \cos x \geq 0$ limits the answers to $x = \boxed{0°, 90°, 135°, \text{ or } 315°}$.

38. **Solution 1:** Draw altitude $\overline{BH}$ and let $x = BH$ and $y = CH$ (see Fig. 4.4). Then $x^2 + y^2 = 169$ and $x^2 + (21 - y)^2 = 400$; subtracting gives $441 - 42y = 231$, so $y = 5$ and $x = 12$. $\cos A = \frac{x}{21-y} = \frac{4}{5}$ and $\cos B = \frac{y}{13} = \frac{5}{13}$, and

$$\cos B = \cos(m\angle CBH + m\angle ABH)$$

$$= \cos(m\angle CBH)\cos(m\angle ABH)$$

$$- \sin(m\angle CBH)\sin(m\angle ABH)$$

$$= \frac{12}{13}\frac{3}{5} - \frac{5}{13}\frac{4}{5}$$

$$= \frac{16}{65},$$

so $\cos A + \cos B + \cos C = \frac{4}{5} + \frac{16}{65} + \frac{5}{13} = \boxed{\frac{93}{65}}$.

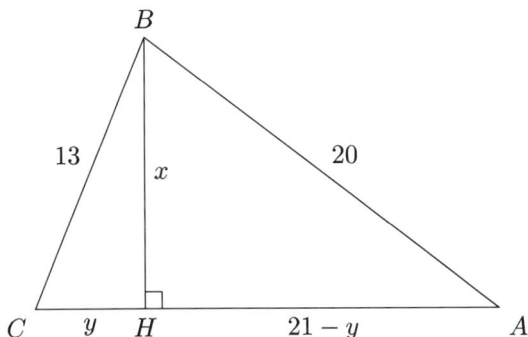

Figure 4.4

Solution 2: By the law of cosines on all three vertices of the triangle,

$$\cos C + \cos B + \cos A = \frac{a^2 + b^2 - c^2}{2ab} + \frac{a^2 - b^2 + c^2}{2ac}$$
$$+ \frac{-a^2 + b^2 + c^2}{2bc}$$
$$= \frac{210}{2 \cdot 21 \cdot 13} + \frac{128}{2 \cdot 13 \cdot 20} + \frac{672}{2 \cdot 21 \cdot 20}$$
$$= \frac{5}{13} + \frac{16}{65} + \frac{4}{5}$$
$$= \boxed{\frac{93}{65}}.$$

39. The sum of the three angles, which must be 180°, is three times the middle angle, so the middle angle is 60°, and that is in fact a sufficient condition for the angles to be in an arithmetic progression. Depending on which side is opposite the 60° angle, the law of cosines can lead to one of three equations:

   - $3^2 = 7^2 + x^2 - 2 \cdot 7 \cdot x \cdot \cos 60°$ or $x^2 - 7x + 40 = 0$, which has no real solutions,
   - $7^2 = 3^2 + x^2 - 2 \cdot 3 \cdot x \cdot \cos 60°$ or $x^2 - 3x - 40 = 0$, which has the positive solution of 8, or

- $x^2 = 7^2 + 3^2 - 2 \cdot 7 \cdot 3 \cdot \cos 60°$ or $x^2 = 37$, which has the positive solution of $\sqrt{37}$.

Thus, the possible solutions are $x = \boxed{\sqrt{37} \text{ or } 8}$.

Level 3

40. Let x be the quantity and multiply by $16 \sin 20°$:

$$16 \sin 20° \cdot x = 16 \sin 20° \cos 20° \cos 40° \cos 60° \cos 80°$$
$$= 8 \sin 20° \cos 20° \cos 40° \cos 80°$$
$$= 4 \sin 40° \cos 40° \cos 80°$$
$$= 2 \sin 80° \cos 80°$$
$$= \sin 160°$$
$$= \sin 20°.$$

So $x = \boxed{\frac{1}{16}}$.

41.

$$\cos 10° \cdot (\cos 20° - 2 \sin^2 10°) = \cos 10° \cos 20° - 2 \cos 10° \sin^2 10°$$
$$= \cos 10° \cos 20° - \sin 10° \sin 20°$$
$$= \cos(10° + 20°)$$
$$= \boxed{\frac{\sqrt{3}}{2}}.$$

42. Note that rotating P and C clockwise by $60°$ around B maps them onto A and Q (see Fig. 4.5), and therefore $PC = AQ$; similarly, rotating around A gives $PC = BR$. By the law of cosines, $\cos(m\angle ABC) = \frac{7^2 - 3^2 - 8^2}{-2 \cdot 3 \cdot 8} = \frac{1}{2}$, so $m\angle ABC = 60°$ and

$m\angle PBC = 120°$. Then $PC^2 = PB^2 + BC^2 - 2 \cdot PB \cdot BC \cdot \cos 120° = 97$, so $PC = QA = RB = \sqrt{97}$ and their sum is $\boxed{3\sqrt{97}}$.

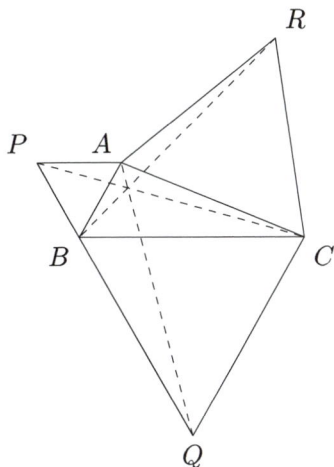

Figure 4.5

43. The medians of a triangle form another triangle whose area is $\frac{3}{4}$ times that of the original triangle. So, if x is the third median, we want a triangle with sides 5, 6, and x whose area is 9. Let θ be the angle between the sides of length 5 and 6 in the new triangle. Then $\frac{5 \cdot 6 \cdot \sin \theta}{2} = 9$, or $\sin \theta = \frac{3}{5}$, so $\cos \theta = \pm\frac{4}{5}$. Then the law of cosines gives $x^2 = 5^2 + 6^2 \pm 2 \cdot 5 \cdot 6 \cdot \frac{4}{5}$, so $x = \boxed{\sqrt{13} \text{ or } \sqrt{109}}$.

44. **Solution 1:** The cosine double angle formulas give $\cos 72° = 2\cos^2 36° - 1$ and $\cos 36° = 1 - 2\sin^2 18°$. Adding these together gives

$$
\begin{aligned}
\cos 72° + \cos 36° &= 2\left(\cos^2 36° - \sin^2 18°\right) \\
&= 2\left(\cos^2 36° - \cos^2 72°\right) \\
&= 2(\cos 36° - \cos 72°)(\cos 36° + \cos 72°) \\
\boxed{-\frac{1}{2}} &= \cos 72° - \cos 36°.
\end{aligned}
$$

Solution 2: As we have seen in problem 23 in the Geometry chapter, a triangle whose angles are 36°, 72°, and

72° and whose equal sides have length 1 looks like this (Fig. 4.6): from which we can read off $\cos 72° = \frac{NB}{AB} = \frac{-1+\sqrt{5}}{4}$

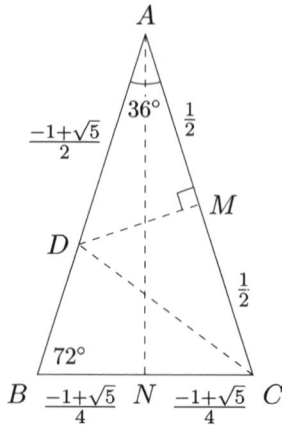

Figure 4.6

and $\cos 36° = \frac{MA}{MD} = \frac{\frac{1}{2}}{\frac{-1+\sqrt{5}}{2}} = \frac{1+\sqrt{5}}{4}$, so $\cos 72° - \cos 36° = \boxed{-\frac{1}{2}}$.

45. Draw the perpendicular from D to $\overline{AB}$, ending at E, and draw the perpendicular from C to $\overline{DE}$, ending at F (see Fig. 4.7). Then

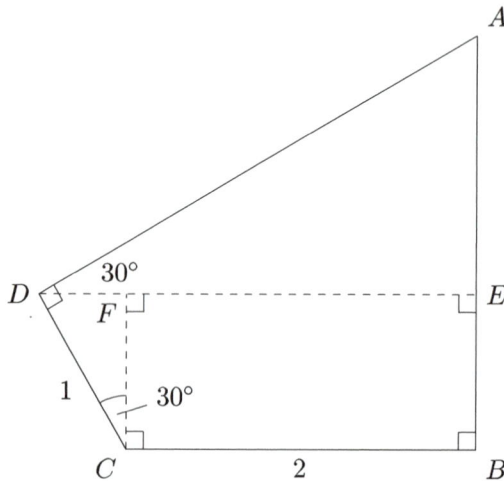

Figure 4.7

$\overline{DE}\|\overline{BC}$, $\overline{BC} \perp \overline{CF}$, $m\angle FCD = m\angle BCD - m\angle BCF = 30°$, $m\angle FDC = 90° - m\angle FCD = 60°$, and $m\angle EDA = m\angle CDA - m\angle FDC = 30°$. So $EFCB$ is a rectangle and $\triangle FDC$ and $\triangle EAD$ are $30 - 60 - 90$ triangles, which means $EB = FC = \frac{\sqrt{3}}{2}CD = \frac{\sqrt{3}}{2}$, and $EA = \frac{ED}{\sqrt{3}} = \frac{EF+FD}{\sqrt{3}} = \frac{5}{2\sqrt{3}} = \frac{5\sqrt{3}}{6}$. Finally, $AB = EA+EB = \boxed{\frac{4\sqrt{3}}{3}}$.

46. Let $s = \sin x$ and $c = \cos x$. In general, $a^3 + b^3 = (a+b)(a^2 - ab + b^2)$, so

$$\frac{2}{3} = s^6 + c^6 = \cancel{(s^2 + c^2)}(s^4 - s^2 c^2 + c^4)$$
$$= (s^2 + c^2)^2 - 3s^2 c^2$$
$$= 1 - 3s^2 c^2$$
$$s^2 c^2 = \frac{1}{9}$$
$$sc = \pm\frac{1}{3}.$$

Then $\sin 2x = 2sc = \pm\frac{2}{3}$, but since $x \in [0°, 90°]$, $2x \in [0°, 180°]$, so $\sin 2x \geq 0$ and $\sin 2x = \boxed{\frac{2}{3}}$.

47. Draw the radius from the center of the circle to each vertex of the polygon, which divides the polygon into n congruent triangles. Each of them has two sides of length r and an angle between them of $\frac{360°}{n}$, so it has area $\frac{r^2}{2}\sin\frac{360°}{n}$, and the total area of the polygon is $\frac{nr^2}{2}\sin\frac{360°}{n}$. Thus we want to find all n such that $f(n) = \frac{n}{2}\sin\frac{360°}{n}$ is an integer. Note that the area of the polygon is bounded above by the area of the circle, which is πr^2, so $f(n)$ is bounded by π, and the only possible integral values it can take are 1, 2, and 3. Further, $f(n)$ is a strictly increasing function of n, so it can only take each value at most once. You can check that $f(4) = 2$ and $f(12) = 3$, and that $f(n)$ can never be 1. Thus the possible values are $n = \boxed{4 \text{ or } 12}$.

48. **Solution 1:** The law of sines shows (as in the appendix) that $\frac{a}{\sin A} = 2R$ or $\sin A = \frac{a}{2R}$ where R is the radius of the circumcircle of the triangle. Using this and the law of cosines, we get

$$\sin A = \frac{a}{2R}$$

$$\cos A = \frac{b^2 + c^2 - a^2}{2bc}$$

$$\cot A = \frac{\cos A}{\sin A} = \frac{R \cdot (b^2 + c^2 - a^2)}{abc}.$$

Similarly,

$$\cot B = \frac{R \cdot (a^2 + c^2 - b^2)}{abc}$$

$$\cot C = \frac{R \cdot (a^2 + b^2 - c^2)}{abc}.$$

Since a^2, b^2 and c^2 form an arithmetic progression, $b^2 = \frac{a^2 + c^2}{2}$. Using this, when we average $\cot A + \cot C$ and simplify, we get $\frac{\cot A + \cot C}{2} = \frac{Rb^2}{abc}$, and using $2b^2 = a^2 + c^2$, we find $\cot B = \frac{R(b^2)}{abc}$. Thus, $\cot B = \frac{\cot A + \cot C}{2} = \boxed{\frac{3}{2}}$.

Solution 2: Since we may scale and move the triangle arbitrarily without breaking the conditions on it, give the vertices coordinates $A(3,0)$, $B(x,y)$, and $C(0,0)$ (see Fig. 4.8). Then $\cot A = \frac{3-x}{y}$ and $\cot C = \frac{x}{y}$, so $3 = \cot A + \cot C = \frac{3}{y}$ and $y = 1$. The arithmetic progression implies that $2b^2 = a^2 + c^2$, or

$$18 = (x^2 + 1) + ((3 - x)^2 + 1)$$
$$= 2x^2 - 6x + 11$$
$$0 = 2x^2 - 6x - 7.$$

The two different roots of this equation correspond to the sequence being either increasing or decreasing from a^2 to c^2. The sum of the roots is $-\frac{-6}{2} = 3$, which means that the two choices for x result in congruent triangles that are just reflected across the perpendicular bisector of $\overline{AC}$, so they're equivalent as far as we're concerned. Note also that the two roots are $\cot A$ and $\cot C$ and that their product is $-\frac{7}{2}$. Finally,

$$\cot B = -\cot(180° - B) = -\cot(A + C) = -\frac{\cot A \cot C - 1}{\cot A + \cot C}$$

$$= -\frac{-\frac{7}{2} - 1}{3} = \boxed{\frac{3}{2}}.$$

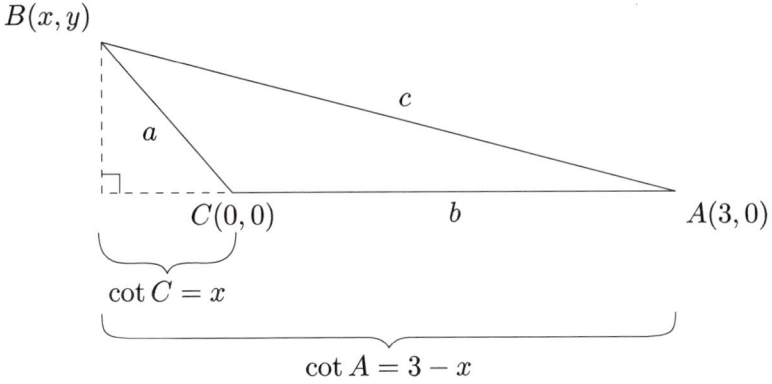

Figure 4.8

Chapter 5

Logarithms

Questions

Level 1

1. Express $\frac{\log 5}{\log \frac{1}{5}}$ in simplest form.

2. If $\log_{10} x = 1.5421$, express $10^{3.5421}$ in terms of x.

3. Compute $\log_2 (\log_{16} 4)^{\log_5 125}$.

4. Compute the numerical value of $\log_{(3^{\log_3 9})} 81$.

5. Compute the numerical value of $2(\log_2 2^2)^2$.

6. If $\log(\log(\log(\log x))) = 0$, where the base of each logarithm is 10, then $x = 10^k$. Find the positive integer k.

7. Find all real x for which $\log_2 \left|1 + \frac{9}{x^2}\right| = 1$.

8. Compute $3^{\log_9 27}$.

9. Compute $2^{\log_3 5} - 5^{\log_3 2}$.

10. If $\log_{10} x = 2 - 2\log_{10} 2$, compute x.

11. Compute $\log_{\sqrt{2}}(\sqrt{2}^{\sqrt{2}})^{\sqrt{2}}$.

12. Find all real x such that $\log_7 5 - \log_7 4 = \log_7 10 - \log_7 x$.

13. Compute the numerical value of x if $x = \log_{3x} 36$.

14. Find all real x which satisfy the equation $\log_{10}\left(x^2 + 5x - 50\right) = 2$.

15. If $a = \log_4 32$ and $b = \log_8 16$, compute the ratio $a : b$ in simplest form.

16. If $\log_4 \log_3 \log_2 x = 1$, the real number x can be expressed as 2^k, where k is a positive real number. Compute k.

17. Compute the numerical value of $5^{\log_{25}\left(7 - \frac{3}{4}\right)}$.

18. Find all real x such that $2^{\log_4 x} = 16$.

19. If $\log_7 x = 0.47$, compute $\log_{49} x^2$.

20. Given $a = \log_{11} 2$, $b = \log_{11} 3$, and $c = \log_{11} 10$, express $\log_{10} 36$ in simplest form in terms of a, b, and c.

Level 2

21. Compute the value of y if $(\log_3 x)(\log_x 2x)(\log_{2x} y) = \log_x(x^2)$.

22. If the geometric mean of $\log_2 17$, $\log_8 17$, and $\log_{512} 17$ is $\log_k 17$, compute k.

23. If $(\log_b a)(\log_c b)(\log_c a) = 25$ and $\frac{a^2}{c^2} = c^k$, compute the sum of all possible values of k.

24. Find both possible values of a if $\log x = a \log y$ and $x^4 - \left(\frac{x}{y}\right)^3 - xy^4 + y = 0$, where x and y are not 1.

25. If $\log_{10}\left(\frac{1}{\sin x}\right) + \log_{10}\left(\frac{1}{\cos x}\right) = 2$ and $\log_{10}(\sin x + \cos x) = \log_{10}\sqrt{\frac{p}{q}}$, where p and q are relatively prime, compute $p + q$.

26. If $\log_5 \sin x = -\frac{1}{2}$, compute the numerical value of $\cos^2 x$.

27. Find all real x for which there exists some integer $k > 1$ such that $(\log_k x)(\log_6 k) = \frac{5}{2}$.

28. If $2 \cdot 4^x + 6^x = 9^x$ and $x = \log_{\frac{2}{3}} a$, find the numerical value of a.

29. Compute $\log_{\frac{1}{8}} \sin 4350°$.

30. Find all real x such that $\log_x 2 + \log_2 x = \frac{5}{2}$.

31. If $m = \log_{10} 2$, express $\log_4 5$ in terms of m.

32. If $\log_2 3^4 \cdot \log_3 4^5 \cdot \log_4 5^6 \cdot \ldots \cdot \log_{63} 64^{65} = x!$ compute x.

33. Find all positive real numbers x such that $(\log_{10} x^2)^2 = \log_{10} 10000$.

34. If $b = \log_8 12$, express $\log_8 3$ in terms of b.

35. Compute the numerical value of $\log_{10}\frac{1}{2} + \log_{10}\frac{2}{3} + \log_{10}\frac{2}{3} + \cdots + \log_{10}\frac{99}{100}$.

36. If x is a positive real number, $a = \log_3 \log_{27} x$, and $b = \log_{27} \log_3 x$, express a explicitly in terms of b.

37. If $16(\log x)^2 + 9(\log y)^2 = 24 \cdot \log x \cdot \log y$, find y in terms of x.

38. Let f be a function of x only and let g be a function of y only. If $z = x + xy + y$, determine some g such that $\log f(x) + \log g(y) = \log(1 + z)$.

39. Compute $\log \tan 1° + \log \tan 2° + \log \tan 3° + \cdots + \log \tan 90°$.

40. Compute all values of x such that $x \in [0, 2\pi]$ and $(\log_{10}(4x - 10))^{4\sin x + 2} = 1$.

Level 3

41. Compute the value of y that satisfies $\log_5 y = (-2\log_5 \sqrt{2})(1 + \log_4 144)$.

42. Compute $\log_{1024} \frac{(\sqrt{3}+i)^{1991}}{\sqrt{3}-i}$.

43. Compute all values of x that satisfy $(\log_4 x^2 - 4)^3 + (\log_8 x^3 - 8)^3 = (\log_2 x^2 - 12)^3$.

44. If $x = \frac{\pi}{9}$, compute $\frac{\log \cos 2x - \log(1 + \sin 2x)}{\log(\cot x + 1) - \log(\cot x - 1)}$.

Answers

Level 1

1. $\frac{\log 5}{\log \frac{1}{5}} = \frac{\log 5}{\log 5^{-1}} = \frac{\log 5}{-\log 5} = \boxed{-1}$.

2. If $\log_{10} x = 1.5421$, then $10^{1.5421} = x$ and $10^{3.5421} = 10^2 \cdot 10^{1.5421} = 10^2 x = \boxed{100x}$.

3. $\log_5 125 = 3$ and $\log_{16} 4 = \frac{1}{2}$, so the expression simplifies to $\log_2 \left(\frac{1}{2}\right)^3 = \log_2 2^{-3} = \boxed{-3}$.

4. Since $3^{\log_3 9} = 9$, we want $\log_9 81 = \log_9 9^2 = \boxed{2}$.

5. $2\left(\log_2 2^2\right)^2 = 2(2)^2 = \boxed{8}$.

6. Note that $k = \log_{10} x$ by definition. We have $\log_{10} \log_{10} \log_{10} x = 10^0 = 1$, so $\log_{10} \log_{10} x = 10^1 = 10$, so $\log_{10} x = \boxed{10^{10}}$.

7. Since $1 + \frac{9}{x^2}$ is positive, $\left|1 + \frac{9}{x^2}\right| = 1 + \frac{9}{x^2}$. So we want $1 + \frac{9}{x^2} = 2$, or $x = \boxed{\pm 3}$.

8. $27 = 3^3 = 9^{\frac{3}{2}}$, so $\log_9 27 = \frac{3}{2}$, and we want $3^{\frac{3}{2}} = \boxed{3\sqrt{3}}$.

9. **Solution 1:** $2^{\log_3 5} = (5^{\log_5 2})^{\log_3 5} = 5^{\log_5 2 \log_3 5} = 5^{\log_3 2}$, so the difference is $\boxed{0}$.

   Solution 2: $2^{\log_3 5} - 5^{\log_3 2} = (3^{\log_3 2})^{\log_3 5} - (3^{\log_3 5})^{\log_3 2} = 3^{\log_3 5 \log_3 2} - 3^{\log_3 5 \log_3 2} = \boxed{0}$.

10. $\log_{10} x = \log_{10} 100 - \log_{10} 4 = \log_{10} \frac{100}{4} = \log_{10} 25$, so $x = \boxed{25}$.

11. $\log_{\sqrt{2}} \left(\sqrt{2}^{\sqrt{2}}\right)^{\sqrt{2}} = \log_{\sqrt{2}}(\sqrt{2})^2 = \boxed{2}$.

12. The given equation is equivalent to $\log_7 \frac{5}{4} = \log_7 \frac{10}{x}$, so $\frac{5}{4} = \frac{10}{x}$ and $x = \boxed{8}$.

13. By inspection, $x = 2$ satisfies the equation. (In a bit more detail: the only "nice" values for $3x$ could be 1, 36, or $\sqrt{36} = 6$, and checking them leads to $3x = 6$.) Meanwhile, the equation is equivalent to $36 = (3x)^x$; $x < 2$ implies $(3x)^x < 36$ and $x > 2$ implies $(3x)^x > 36$, so $x = \boxed{2}$ is the only solution.

14. Raising 10 to the power of both sides gives $x^2 + 5x - 50 = 100$ or $x^2 + 5x - 150 = 0$, which has solutions $x = \boxed{10 \text{ or } -15}$.

15. $a = \log_{2^2} 2^5 = \frac{5}{2}$ and $b = \log_{2^3} 2^4 = \frac{4}{3}$, so $a : b = \frac{5}{2} : \frac{4}{3} = \boxed{15 : 8}$.

16. $k = \log_2 x$ by definition, and we have $\log_3 \log_2 x = 4^1 = 4$ and $\log_2 x = 3^4 = \boxed{81}$.

17. $5^{\log_{25}\left(7-\frac{3}{4}\right)} = 5^{\log_{25} \frac{25}{4}} = 25^{\frac{1}{2} \log_{25} \frac{25}{4}} = 25^{\log_{25} \frac{5}{2}} = \boxed{\frac{5}{2}}$.

18. $2^{\log_4 x} = 2^4$, so $\log_4 x = 4$ and $x = 4^4 = \boxed{256}$.

19. The change of base formula gives $\log_{49} x^2 = \frac{\log_7 x^2}{\log_7 49} = \frac{2\log_7 x}{2} = \boxed{0.47}$.

20. $\log_{10} 36 = \frac{\log_{11} 36}{\log_{11} 10} = \frac{2\log_{11} 2 + 2\log_{11} 3}{\log_{11} 10} = \boxed{\frac{2a+2b}{c}}$.

Level 2

21. Since $\log_b a = \frac{\log a}{\log b}$, $2 = \frac{\log x \; \log 2x \; \log y}{\log 3 \; \log x \; \log 2x} = \frac{\log y}{\log 3}$. So $y = 3^2 = \boxed{9}$.

22. Note that $8 = 2^3$ and $512 = 2^9$, so the geometric mean is

$$\sqrt[3]{\frac{\log 17}{\log 2}\frac{\log 17}{\log 8}\frac{\log 17}{\log 512}} = \sqrt[3]{\frac{\log 17}{\log 2}\frac{\log 17}{3\log 2}\frac{\log 17}{9\log 2}}$$

$$= \frac{\log 17}{\log 2}\sqrt[3]{1 \cdot \frac{1}{3} \cdot \frac{1}{9}}$$

$$= \frac{\log 17}{3\log 2}$$

$$= \log_8 17,$$

so $k = \boxed{8}$.

23. $\log_b a \log_c b = \log_c a$, so we have $(\log_c a)^2 = 25$ and $\log_c a = \pm 5$. Meanwhile, $\frac{a^2}{c^2} = c^k$ means $a = c^{\frac{k+2}{2}}$, so $\log_c a = \frac{k+2}{2} = \pm 5$ and $k = -2 \pm 10$, so the sum of the two possible values of k is $\boxed{-4}$.

24. Note that $a = \log_y x$. Factor the equation to get $0 = x^3 \cdot \left(x - \frac{1}{y^3}\right) - y^4\left(x - \frac{1}{y^3}\right) = (x^3 - y^4)(x - y^{-3})$, so $x = y^{\frac{4}{3}}$ or $x = y^{-3}$. Thus $\log_y x = \boxed{\frac{4}{3} \text{ or } -3}$.

25. The first equation gives $\log_{10}\left(\frac{1}{\sin x \cos x}\right) = 2$ or $\log_{10}(\sin x \cos x) = -2$, so $\sin x \cos x = 10^{-2} = \frac{1}{100}$. The second equation gives $\sin x +$

$\cos x = \sqrt{\frac{p}{q}}$, so $\frac{p}{q} = (\sin x + \cos x)^2 = 1 + 2\sin x \cos x = 1 + \frac{2}{100} = \frac{51}{50}$. Thus $p = 51$ and $q = 50$, so $p + q = \boxed{101}$.

26. $\sin x = 5^{-\frac{1}{2}} = \sqrt{5}$ and $\cos^2 x = 1 - \sin^2 x = 1 - \frac{1}{5} = \boxed{\frac{4}{5}}$.

27. The change of base formula gives $\frac{5}{2} = \log_k x \log_5 k = \frac{\log x \, \log k}{\log k \, \log 5} = \log_5 x$ (so, in fact, k is irrelevant as long as all the logs are defined). Thus $x = 5^{\frac{5}{2}} = \boxed{25\sqrt{5}}$.

28. Divide by 9^x to get $1 = 2\left(\frac{4}{9}\right)^x + \left(\frac{2}{3}\right)^x = 2\left(\frac{2}{3}\right)^{2x} + \left(\frac{2}{3}\right)^x$. Substitute $a = \left(\frac{2}{3}\right)^x$ to get $2a^2 + a = 1$, which has positive root $\boxed{\frac{1}{2}}$.

29. $\sin(4350°) = \sin(12 \cdot 360° + 30°) = \sin 30° = \frac{1}{2}$, so we want $\log_{\left(\frac{1}{2}\right)^3} \frac{1}{2} = \boxed{\frac{1}{3}}$.

30. Let $u = \log_x 2$, and so $\log_2 x = \frac{1}{u}$. Thus we have $u + \frac{1}{u} = \frac{5}{2}$, which has solutions 2 and $\frac{1}{2}$. So $x = 2^2$ or $2^{\frac{1}{2}}$, i.e., $x = \boxed{4 \text{ or } \sqrt{2}}$.

31. **Solution 1:**

$$\log_{10} 2 = m$$

$$\log_{10} 4 = 2m$$

$$\log_4 10 = \frac{1}{2m}$$

$$\log_4(5) = \log_4\left(\frac{10}{2}\right) = \log_4 10 - \log_4 2$$

$$\log_4 5 = \frac{1}{2m} - \frac{1}{2} = \boxed{\frac{1-m}{2m}}.$$

Solution 2: Let $y = \log_4 5$, so $4^y = 5 = \frac{10}{2}$. Take $10^m = 2$ and raise both sides to the $2y$ power to get $10^{2my} = 2^{2y} = 4^y = \frac{10}{2}$, so $2my = \log_{10} 10 - \log_{10} 2 = 1 - m$, and $y = \boxed{\frac{1-m}{2m}}$.

32.

$$x! = 4 \cdot 5 \cdot 6 \cdot \ldots \cdot 65 \cdot \log_2 3 \cdot \log_3 4 \cdot \log_4 5 \cdot \ldots \cdot \log_{63} 64.$$

So

$$x! = 4 \cdot 5 \cdot 6 \cdot \ldots \cdot 65 \cdot \frac{\log 3}{\log 2} \cdot \frac{\log 4}{\log 3} \cdot \frac{\log 5}{\log 4} \cdot \ldots \cdot \frac{\log 64}{\log 63}.$$

So now

$$x! = 4 \cdot 5 \cdot 6 \cdot \ldots \cdot 65 \cdot \frac{\log 64}{\log 2} = 4 \cdot 5 \cdot 6 \cdot \ldots \cdot 65 \cdot \frac{\log 2^6}{\log 2}.$$

Simplifying we get

$$x! = 4 \cdot 5 \cdot 6 \cdot \ldots \cdot 65 \cdot 6 = 1 \cdot 2 \cdot 3 \cdot 4 \cdot 5 \cdot 6 \cdot \ldots \cdot 65, \text{ so } x = \boxed{65}.$$

33. We have $(\log_{10} x^2)^2 = 4$, so $\log_{10} x^2 = \pm 2$ and (since x must be positive) $\log_{10} x = \pm 1$, so $x = \boxed{10 \text{ or } \frac{1}{10}}$.

34. $\log_8 3 = \log_8 \frac{12}{4} = \log_8 12 - \log_8 4 = b - \log_{2^3} 2^2 = \boxed{b - \frac{2}{3}}$.

35.

$$\log_{10} \frac{1}{2} + \log_{10} \frac{2}{3} + \log_{10} \frac{3}{4} + \cdots + \log_{10} \frac{98}{99} + \log_{10} \frac{99}{100}$$

$$= \log_{10} \left(\frac{1}{\cancel{2}} \cdot \frac{\cancel{2}}{\cancel{3}} \cdot \frac{\cancel{3}}{\cancel{4}} \cdot \ldots \cdot \frac{\cancel{98}}{\cancel{99}} \frac{\cancel{99}}{100} \right)$$

$$= \log_{10} \frac{1}{100}$$

$$= \boxed{-2}.$$

36. **Solution 1:** The definition of a gives $x = 27^{3^a} = (3^3)^{(3^a)} = 3^{3 \cdot 3^a} = 3^{3^{a+1}}$. On the other hand, the definition of b gives $x = 3^{27^b} = 3^{(3^3)^b} = 3^{3^{3b}}$. Thus $3b = a + 1$, or $a = \boxed{3b - 1}$.
Solution 2: For any y, $\log_{27} y = \log_{3^3} y = \frac{1}{3} \log_3 y$, so

$$a = \log_3 \left(\frac{1}{3} \log_3 x \right)$$

$$= (\log_3 \log_3 x) - 1$$

$$b = \frac{1}{3} \log_3 \log_3 x$$

$$a = \boxed{3b - 1}.$$

37. Moving all the terms to one side and factoring gives

$$(4 \log x - 3 \log y)^2 = 0$$

$$4 \log x = 3 \log y$$

$$\log_x y = \frac{\log y}{\log x} = \frac{4}{3},$$

so $y = \boxed{x^{\frac{4}{3}}}$.

38. $\log(1+z) = \log(1+x+y+xy) = \log(1+x)(1+y) = \log(1+x)+\log(1+y)$. Thus $g(y) = \boxed{1 + y}$. (It could also be any positive multiple of $1+y$, since we can write $\log(1+z) = (\log(1+x)-c)+(\log(1+y)+c) = \log(e^{-c}(1 + x)) + \log\left(e^c(1 + y)\right)$.)

39. Since $\log \tan(90° - x) = \log \frac{1}{\tan x} = -\log \tan x$, all the terms cancel out except $\log \tan 45° = \log 1 = \boxed{0}$.

40. Let $a = \log_{10}(4x - 10)$ and $b = 4\sin x + 2$; $a^b = 1$ if and only if one of the following conditions is true:

- $a = 1$, which implies $4x - 10 = 10$ and so $x = 5$;
- $b = 0$, which implies $\sin x = -\frac{1}{2}$ and so $x = \frac{7\pi}{6}$ or $\frac{11\pi}{6}$; or
- $a = -1$ and b is an even integer, which is impossible, because $a = -1$ implies $4x - 10 = 0.1$ and so $x = 2.525$, but then $4\sin x - 2$ is not an even integer (it is not even an integer!).

Thus the possible solutions are $x = \boxed{5, \frac{7\pi}{6}, \text{ or } \frac{11\pi}{6}}$.

Level 3

41.

$$
\begin{aligned}
\log_5 y &= (-2\log_5 \sqrt{2})(1 + \log_4 144) \\
&= (-\log_5 2)(1 + \log_4 144) \\
&= -\log_5 2 - \log_5 2 \cdot \log_4 144 \quad \text{(use the change of base} \\
&= -\frac{\log 2}{\log 5} - \frac{\log 2}{\log 5} \cdot \frac{\log 144}{2\log 2} \quad \text{formula to convert to} \\
&\qquad\qquad\qquad\qquad\qquad\qquad\quad \text{base 10)} \\
&= -\frac{\log 2}{\log 5} - \frac{\log 2}{\log 5} \cdot \frac{2\log 12}{2\log 2} \\
&= -\frac{\log 2}{\log 5} - \frac{\log 12}{\log 5} = -\frac{\log 24}{\log 5} = \log_5 \frac{1}{24}.
\end{aligned}
$$

So $y = \boxed{\frac{1}{24}}$.

42. $\sqrt{3} + i = 2e^{\frac{i\pi}{6}}$ and $\sqrt{3} - i = 2e^{-\frac{i\pi}{6}}$, so

$$
\frac{\left(\sqrt{3} + i\right)^{1991}}{\sqrt{3} - i} = \frac{2^{1991}e^{\frac{1991\pi i}{6}}}{2e^{-\frac{i\pi}{6}}} = 2^{1990}e^{\frac{1992\pi i}{6}} = 2^{1990},
$$

with the last equality arising because 1992 is a multiple of 12, so that $\frac{1992\pi}{6}$ is a multiple of 2π and $e^{\frac{1992\pi i}{6}} = 1$. Finally, $1024 = 2^{10}$, so we want $\log_{2^{10}} 2^{1990} = \boxed{199}$.

43. $\log_4 x^2 = \frac{\log x^2}{\log 4} = \frac{2\log x}{2\log 2} = \log_2 x$, $\log_8 x^3 = \frac{\log x^3}{\log 8} = \frac{3\log x}{3\log 2} = \log_2 x$ and $\log_2 x^2 = 2\log_2 x$, so $(\log_2 x - 4)^3 + (\log_2 x - 8)^3 =$

$(2\log_2 x - 12)^3$. If we let $a = \log_2 x - 4$ and $b = \log_2 x - 8$, this becomes $a^3 + b^3 = (a + b)^3$, or $0 = 3a^2b + 3b^2a = 3ab(a + b)$. This can be satisfied if $a = 0$, which implies $x = 16$; $b = 0$, which implies $x = 256$; or $a + b = 0$, which implies $2\log_2 x - 12 = 0$ and so $x = 64$. So the solutions are $x = \boxed{16, 64, \text{ or } 256}$.

44. This expression is equivalent to $\dfrac{\log \frac{\cos 2x}{1+\sin 2x}}{\log \frac{\cot x+1}{\cot x-1}}$.

$$\frac{\cos 2x}{1 + \sin 2x} = \frac{\cos^2 x - \sin^2 x}{\sin^2 x + \cos^2 x + 2\sin x \cos x}$$

$$= \frac{(\cos x - \sin x)(\cos x + \sin x)}{(\sin x + \cos x)^2}$$

$$= \frac{\cos x - \sin x}{\cos x + \sin x} = \frac{\frac{\cos x}{\sin x} - 1}{\frac{\cos x}{\sin x} + 1}$$

$$= \frac{\cot x - 1}{\cot x + 1} = \frac{1}{\frac{\cot x+1}{\cot x-1}}$$

so that the two logs are negatives of each other and their quotient is $\boxed{-1}$ (independent of x, as long as there are no divisions by zero, which there are not for $x = \frac{\pi}{9}$).

Chapter 6

Counting

Questions

Level 1

1. How many distinct points do the graphs of $x^2 + y^2 = 16$ and $y = x^2 - 4$ have in common?

2. For how many real values of x does the expression $\frac{2}{1 - \frac{1}{x-3}}$ have no value?

3. A committee of 24 senators votes on a bill, with each senator voting either "yes" or "no." The bill will pass if the number of "yes" votes is at least twice the number of "no" votes. Ten of the senators vote "no." At least how many of these senators would have to change their minds for the bill to pass?

4. How many ordered triples of positive integers (x, y, z) satisfy $x < y < z$ and $xyz = 105$?

5. Cathy and Jamie are both invited to a dinner party. There will be a total of six people at the party, seated around a round table. Cathy and Jamie cannot sit next to each other without causing a scene. All the other guests, however, are compatible with each other and with Cathy and Jamie. Compute the number of different ways in which the guests can be seated without causing a scene. (Two arrangements are considered different if and only if any person has a different neighbor to the right in the two arrangements.)

6. The number N is represented by a two-digit base-ten numeral. If the two digits are reversed and the resulting number is added to N, the result is a perfect square. How many possible values of N are there?

7. A student wrote the algebraic expression $(((x - 7) \; \hat{} \; ((x + 3) - (x + 4) \; \hat{} \; 2) - (x - 3) \; \hat{} \; (x \; \hat{} \; (2x) + 1)(x^2 + 1)$, in which the symbol $\hat{}$ indicates raising to a power. Determine how many necessary closing parentheses the student left out.

8. Find the number of positive integral solutions to $2x + 3y = 1009$.

9. How many ordered pairs of positive integers (a, b) satisfy $a^2 - b^2 = 105$?

10. Henry and Eliza have a reading assignment for English. Their teacher has assigned them four books to read for the month. The friends decide to save time and work by splitting the assignment: Henry will read two books and Eliza the other two. In how many ways can they divide up the four books in this manner?

11. How many natural numbers n are there such that $\frac{2}{5} < \frac{n}{17} < \frac{11}{13}$?

12. An interior diagonal of a polyhedron is a line segment connecting two of the polyhedron's vertices, but not lying entirely on a face of the polyhedron. How many interior diagonals does a cube have?

13. Compute the number of sets of positive integers the sum of whose elements is 7.

14. Two sides of a triangle with nonzero area measure 6 and 11. How many different integral lengths can the third side have?

15. If 20 people are in a room and each one shakes hands with everyone else exactly once, how many handshakes take place?

16. A pizza parlor offers 5 different toppings: pepperoni, mushrooms, broccoli, onions, and meatballs. Slices can come with any number of toppings, even none. If a pizza lover buys a different type of slice every day, how many days will it take to have a slice of every type of pizza?

17. How many five-digit numbers have their digits in decreasing order from left to right with no repetitions?

18. A digital clock gives the time as 6:15 or 12:34, etc. If we ignore the colons, the times represent 3- or 4-digit integers (e.g., 615 or 1234). How many multiples of 3 occur in this fashion during the 12-hour period from noon through 11:59 p.m., inclusive?

Level 2

19. Compute the number of non-congruent right triangles whose sides all have integral lengths and which have one leg of length 40.

20. How many real numbers x satisfy $\log_{3-x}\left(\frac{1}{|x|}\right) = 1$?

21. How many ordered pairs of integers (x, y) satisfy $x^2 = y^2 + 2y + 13$?

22. Compute the number of ways in which four men and four women can be seated at a circular table if men and women must alternate in the arrangement.

23. If $2\sin x + 3\sin 2x = 0$ and $x \in [0, 2\pi]$, compute the number of possible values of x.

24. Melmacian license plates consist of exactly three letters of the standard English alphabet. A plate is valid if and only if the letter Q, if present, is immediately followed by the letter U, and two license plates are considered identical if and only if they contain the same three letters in the same order. How many distinct valid Melmacian license plates are there?

25. How many positive integers n do <u>not</u> satisfy the inequality $n^{\log_{10}\sqrt{n}} > n$?

26. How many integers between 1000 and 10000 inclusive are perfect cubes?

27. How many positive integers n there are such that $n^5 + 3$ is evenly divisible by $n^2 + 1$?

28. How many positive multiples of 8 are made up of the digits 1, 2, 3, 4, and 5, with each digit used at most once?

29. How many natural numbers of the form $1\underline{2}\underline{8}\underline{7}\underline{a}\underline{b}6$, where a and b stand for single digits, are multiples of 72?

30. X is a positive two-digit integer and Y is the two-digit integer formed by reversing the digits of X. How many values of X are there such that $X - Y$ is a perfect square?

31. The 25 teams in the little league are divided into two divisions, A and B. Each team plays against every other team in its division exactly once, and does not play against any teams in the other division. If 36 more games are played in division A than in division B, how many teams are in division A?

32. At a certain high school reunion, everyone shakes hands with everyone else exactly once. If there were 3160 handshakes in total, how many people attended the reunion?

33. A professor can only assign grades of A, B, C, or F in a course containing 12 students. If she decides to give an equal number of each grade, how many ways can she assign the grades?

34. How many distinct terms will there be if $(x+y+z)^{17}$ is algebraically expanded and simplified?

35. In how many ways can 6 different charms be arranged on a circular charm bracelet with no discernible starting point?

36. How many different strings of 4 distinct letters are in alphabetic order from left to right and have one of N, Y, C, M, and L as their third letter?

37. How many sets of two or more consecutive positive integers have a sum of 100?

38. In a rectangular coordinate system, the vertices of triangle $\triangle ABC$ are $A(3,3)$, $B(7,7)$, and $C(12,3)$. How many lattice points (points with integral x and y coordinates) are located on the sides of the triangle?

39. Compute the number of integers between 1 and 2006 inclusive that are not divisible by 3 or 7.

40. For how many positive integers n is $1155 + n^2$ a perfect square?

Level 3

41. How many ordered triples of integers (x, y, z) satisfy $x^2 + y^2 + z^2 < 7$?

42. How many positive integers less than or equal to 100 can be expressed as $\lfloor 2x \rfloor + \lfloor 3x \rfloor$ for some real number x, where $\lfloor z \rfloor$ represents the largest integer less than or equal to z?

43. How many of the numbers

- 16
- 1,156
- 111,556
- 11,115,556
- 1,111,155,556

are perfect squares?

44. How many non-congruent triangles are there with positive area, integral side lengths, and perimeter 15?

Answers

Level 1

1. **Solution 1:** The first graph is a circle centered at the origin with radius 4. The second is the parabola $y = x^2$ shifted down by 4 units

(see Fig. 6.1). Thus there is one intersection (a tangent point) at $(0, -4)$ and two other intersections, for a total of $\boxed{3}$ points.

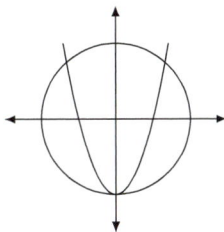

Figure 6.1

Solution 2: Solve algebraically: substituting the second equation into the first gives $16 = x^2 + (x^2 - 4)^2 = x^4 - 7x^2 + 16$, so $x^4 = 7x^2$. This has a double root (tangent point) at $x = 0$, and two more roots, at $x = \pm\sqrt{7}$. Each solution has exactly one corresponding value of y, so there are $\boxed{3}$ intersections.

2. The fraction is undefined when its denominator, $1 - \frac{1}{x-3}$, equals 0, which occurs when $x = 4$. Meanwhile, the denominator itself is undefined when $x - 3 = 0$, or $x = 3$. So there are $\boxed{2}$ values.

3. **Solution 1:** The condition for passing can be restated as the number of "yes" votes needing to be at least $\frac{2}{3}$ of all the senators, or 16. Since 14 senators have already voted "yes," $\boxed{2}$ would need to change their votes.

   Solution 2: If n senators change their votes, then $14 + n$ will be voting "yes" and $10 - n$ will be voting "no." We want $14 + n \geq 2(10 - n)$, which gives $n \geq \boxed{2}$.

4. We have $105 = 3 \cdot 5 \cdot 7$, so one answer is $(3, 5, 7)$. That is the only valid triple whose smallest number is 3, since the only other way to factor $5 \cdot 7$ is $1 \cdot 35$, and $1 < 3$. Also, the smallest number in the triple cannot be bigger than 3, since the next smallest factor of 105 is 5, and $105 < 5^3$. Thus, in any other valid triple, the smallest number must be 1. 105 can be factored into two numbers as $105 = 3 \cdot 35 = 5 \cdot 21 = 7 \cdot 15$, each of which leads to one valid triple, so there are $\boxed{4}$ in total.

5. Two arrangements are considered identical if they are rotations of one another, so it does not matter where Cathy sits. Jamie can take any of the 3 seats that are not next to her. The other guests can take the remaining seats in $4! = 24$ ways, so the total number of arrangements is $3 \cdot 4! = \boxed{72}$.

6. If we write $N = \underline{t}\,\underline{u}$, so that $N = 10t + u$, reversing the digits gives $10u + t$, so $10t + u + 10 + u + t = 11(t + u)$. Since $t, u \leq 9$, we must have $11(t + u) \leq 11 \cdot 18$, so the only way for $11(t + u)$ to be a perfect square is if $t + u = 11$. There are $\boxed{8}$ sets of values that work $(t = 2, \ldots, 9$ and $u = 9, \ldots, 2)$.

7. In a balanced expression, there are as many opening parentheses as closing ones, so the number of missing closing parentheses is the number of opening ones minus the number of closing ones, which is $10 - 8 = \boxed{2}$.

8. Since $2x$ is even and at least 2, $3y \leq 1007$, so $y \leq \lfloor \frac{1007}{3} \rfloor = 335$. But y cannot be even; if it is, $2x$ and $3y$ are both even, but they must sum to an odd number. Any odd value of y leads to a corresponding integer x, so $y = 1, 3, \ldots, 335$ are valid, leading to $\frac{335-1}{2} + 1 = \boxed{168}$ solutions.

9. $105 = (a + b)(a - b)$; since $a + b$ and $a - b$ are both integers, they must equal one of the pairs of positive factors of 105, which are $(105, 1)$, $(35, 3)$, $(21, 5)$, and $(15, 7)$. In any case, $a + b$ must equal the larger one and $a - b$ must equal the smaller one. Since each pair of factors has the same parity, all of them lead to integer solutions, so there are $\boxed{4}$ solutions.

10. We need only count the number of ways in which Eliza can choose two of the books, since Henry must then choose the other two. There are $\binom{4}{2} = \boxed{6}$ ways.

11. If $\frac{2}{5} < \frac{n}{17}$, then $5n > 34$, or $n > 6.8 > 6$. If $\frac{n}{17} < \frac{11}{13}$, then $n < \frac{187}{13} < 15$. So $6 < n < 15$, which leads to $15 - 6 - 1 = \boxed{8}$ solutions.

12. A cube has eight vertices, and for any vertex, only one other vertex does not share an edge or face with it, so each vertex is on exactly one body diagonal. Each edge contains two vertices, so the total number of body diagonals is $\frac{8}{2} = \boxed{4}$.

13. There is one valid set with one element: $\{7\}$. There are three sets with two elements: $\{1, 6\}$, $\{2, 5\}$, and $\{3, 4\}$. There is one set with three elements: $\{1, 2, 4\}$. So there are $\boxed{5}$ sets in total.

14. Let the third side be x. The triangle inequality gives $6 + 11 > x$, $6 + x > 11$, and $x + 11 > 6$. The third inequality is always true, and the first two give $5 < x < 17$, so there are $17 - 5 - 1 = \boxed{11}$ possible values of x.

15. There is exactly one handshake between every unordered pair of people, so the total number is $\binom{20}{2} = \frac{20 \cdot 19}{2} = \boxed{190}$.

16. **Solution 1:** Each topping can either be on a slice or not on the slice, so there are $2^5 = \boxed{32}$ different kinds of slices.

Solution 2: The number of toppings on a slice can be anything between 0 and 5. The number of different kinds of slices with k toppings is $\binom{5}{k}$. Thus the total number of kinds of slices is $\binom{5}{0} + \binom{5}{1} + \binom{5}{2} + \binom{5}{3} + \binom{5}{4} + \binom{5}{5} = 1 + 5 + 10 + 10 + 5 + 1 = \boxed{32}$.

17. Simply pick any five digits without replacement from $\{0, 1, \ldots, 9\}$ and put them in decreasing order. (Note that if the problem asked for digits in increasing order from left to right, we would have to exclude 0.) There are $\binom{10}{5} = \boxed{252}$ ways to do this.

18. The times derived from any given hour (e.g., $1200, \ldots, 1259$) consist of 60 consecutive integers, so there are $\frac{60}{3} = 20$ multiples of 3 among them. There are 12 different values for the hour, so there are $12 \cdot 20 = \boxed{240}$ multiples in total.

Level 2

19. Let b be the length of the other leg and c be the length of the hypotenuse. Thus $40^2 = c^2 - b^2$, or $1600 = (c + b)(c - b)$. Thus $c + b$ is one factor of 1600, and $c - b$ is the inverse factor. Since c is an integer and is the average of these two, they must have the same parity, and so must both be even (since their product is even). Then $\frac{c+b}{2}$ and $\frac{c-b}{2}$ are any two distinct integers with a product of $\frac{1600}{4} = 400$. Since $400 = 2^4 \cdot 5^2$, it has $(4+1) \cdot (2+1) = 15$ factors. Eliminate $\sqrt{400}$ (since it would pair with itself), leaving 14 factors, which separate into $\frac{14}{2} = \boxed{7}$ pairs.

20. Since the base of a logarithm must be positive, $x < 3$. The given equality is equivalent to $\frac{1}{|x|} = 3 - x$. If $x > 0$, this becomes $\frac{1}{x} = 3 - x$ or $x^2 - 3x + 1 = 0$, with solutions $\frac{3 \pm \sqrt{5}}{2}$, which are both less than 3. If $x < 0$, we have $x^2 - 3x - 1 = 0$, with solutions $\frac{3 \pm \sqrt{13}}{2}$. In this case, the larger solution is greater than 3, so we have a total of $\boxed{3}$ solutions.

21. Completing the square gives

$$x^2 = (y+1)^2 + 12$$
$$x^2 - (y+1)^2 = 12$$
$$(x - y - 1)(x + y + 1) = 12$$

So $x - y - 1$ and $x + y + 1$ are two factors of 12. Since x is an integer and is the average of them, they must have the same parity, so they must both be even (since their product is even). Since $12 = 2^2 \cdot 3$, the possible values for them are $(2, 2 \cdot 3)$, $(2 \cdot 3, 2)$, $(-2, -2 \cdot 3)$, and $(-2 \cdot 3, -2)$, so there are $\boxed{4}$ possibilities.

22. We can place one of the women at any position at the table; since it is circular, the position does not matter. Then whether each seat is occupied by a man or a woman is fixed. Thus there are 3! ways to arrange the other women, and 4! ways to arrange the men, for a total of $3! \cdot 4! = \boxed{144}$ ways.

23. Equivalently, $0 = 2 \sin x + 6 \sin x \cos x = 2 \sin x (1 + 3 \cos x)$, so $\sin x = 0$ or $1 + 3 \cos x = 0$. $\sin x = 0$ leads to the solutions 0, π, and 2π, and $1 + 3 \cos x = 0$ leads to two solutions, one in the second quadrant and one in the third quadrant. So there are $\boxed{5}$ solutions in total.

24. A valid license plate either does not contain Q at all, or contains Q followed by U. There are 25 letters besides Q, so there are $25^3 = 15625$ plates not containing Q. If Q is the first letter of a plate, then the second letter must be U and the third letter may be anything besides Q, so there are 25 possible plates; similarly, there are 25 plates with Q as the second letter. There are none with Q as the third letter, since there it cannot be followed by anything. Thus the total number of valid plates is $15625 + 25 + 25 = \boxed{15675}$.

25. Take the logarithm of both sides to get $(\log_{10} n)(\log_{10} \sqrt{n}) > \log_{10} n$. If $n = 1$, both sides are 0 and the inequality is false. Otherwise, we can divide by $\log_{10} n$ to get $\log_{10} \sqrt{n} > 1$, or $n > 100$, so the inequality is not satisfied if $n \in [1, 100]$, so there are $\boxed{100}$ values.

26. We have $10^3 = 1000$ and $21^3 < 10000 < 22^3$, so n^3 is included in the range if $n \in [10, 21]$, which contains $21 - 10 + 1 = \boxed{12}$ values.

27. Long division gives $\frac{n^5 + 3}{n^2 + 1} = n^3 - n + \frac{n+3}{n^2+1}$, so we want $\frac{n+3}{n^2+1}$ to be an integer. If $|n| > 3$, then the absolute value of this expression is less than 1; checking all the remaining values shows that -3, -1, 0, 1, and 2 work, so there are $\boxed{5}$ values.

28. There are no single-digit multiples of 8, since we don't have the digit 8. Any longer multiple must end in either 2 or 4. Checking all the possible two-digit numbers shows that 24 and 32 work. If $a\underline{b}\underline{c}$ is a three-digit multiple, then $\underline{b}\underline{c}$ must be a multiple of 4, so it could be 24 or 32 as before, but also 12 or 52. If $\underline{b}\underline{c}$ is a multiple of 8, then $a\underline{b}\underline{c} - \underline{b}\underline{c} = 100a = 25 \cdot 4a$ must be one as well,

which means that a is even; if instead $\underline{b}\,\underline{c}$ is a multiple of 4 but not 8, then a must be odd. So 24 leads to no three-digit multiples (since it already includes both even digits), 32 leads to one, and 12 and 52 to two each, so there are five three-digit multiples. Now, since 1000 is a multiple of 8, any three-digit multiple $\underline{c}\,\underline{d}\,\underline{e}$ can be extended using the remaining digits into the four- or five-digit multiples $\underline{b}\,\underline{c}\,\underline{d}\,\underline{e}$, $\underline{a}\,\underline{c}\,\underline{d}\,\underline{e}$, $\underline{a}\,\underline{b}\,\underline{c}\,\underline{d}\,\underline{e}$, and $\underline{b}\,\underline{a}\,\underline{c}\,\underline{d}\,\underline{e}$. Thus there are $5 \cdot 4 = 20$ four- or five-digit multiples, for a total of $2 + 5 + 20 = \boxed{27}$.

29. A number is a multiple of 72 if and only if it is a multiple of both 8 and 9. The divisibility test for 9 tells us that the sum of the digits, $1 + 2 + 8 + 7 + a + b + 6 = 24 + a + b$, is a multiple of 9, so $a + b = 3$ or 12. Since the number is divisible by 8, the number formed by the last three digits, $\underline{x}\,\underline{y}\,\underline{6}$, must be divisible by 8. When $y = 9, x = 3$, we get 396, which is not divisible by 8. For $y = 5$, we have $x = 7$ and 756 is not a multiple of 8. For $y = 7$ we have $x = 5$, and for $y = 3$ we have $x = 0$ or $x = 9$. For $y = 1$ we have $x = 2$. Of these, only $(x, y) = (5, 7), (9, 3)$, and $(2, 1)$ work. So there are $\boxed{3}$ pairs that work.

30. Let $X = \underline{a}\,\underline{b} = 10a + b$, so $Y = \underline{b}\,\underline{a} = 10b + a$; since X and Y both have two digits, a and b must be nonzero. We have $X - Y = 9(a - b)$; since 9 is a perfect square, this is a perfect square if and only if $a - b$ is. Thus $a - b = 0, 1, 4$, or 9. If $a - b = c$, with $c \in [0, 9]$, then b can range from 1 up to $9 - c$, so there are $9 - c$ solutions. Thus there are $(9 - 0) + (9 - 1) + (9 - 4) + (9 - 9) = \boxed{22}$ possibilities.

31. Let n be the number of teams in division A.

$$\binom{n}{2} = \binom{25 - n}{2} + 36$$

$$\frac{n(n - 1)}{2} = \frac{(25 - n)(24 - n)}{2} + 36$$

$$n^2 - n = n^2 - 49n + 600 + 72$$

$$48n = 672$$

$$n = \boxed{14}.$$

32. Each pair of people leads to exactly one handshake; there are $\binom{n}{2} = \frac{n(n-1)}{2}$ pairs, so $n(n - 1) = 2 \cdot 3160 = 6320 = 79 \cdot 80$, so $n = \boxed{80}$.

33. There will be $\frac{12}{4} = 3$ students who get each grade. There are $\binom{12}{3}$ ways to assign the As, $\binom{9}{3}$ ways to assign the Bs to the remaining

students, $\binom{6}{3}$ ways to assign the Cs, and $\binom{3}{3} = 1$ way to assign the Fs. The total number of ways is $\frac{12\cdot 11\cdot 10}{3\cdot 2\cdot 1} \cdot \frac{9\cdot 8\cdot 7}{3\cdot 2\cdot 1} \cdot \frac{6\cdot 5\cdot 4}{3\cdot 2\cdot 1} \cdot 1 = \boxed{369600}$.

34. **Solution 1:** Each term in the expansion is $x^a y^b z^c$ multiplied by some constant, where $a+b+c = 17$, and all possible values of a, b, and c are present. This equals the number of ways to arrange 17 stars and 2 separators in a line, which is $\binom{17+2}{2} = \frac{19\cdot 18}{2} = \boxed{171}$.

Solution 2: Use the binomial theorem to expand $(a+b)^{17}$, where $a = x$ and $b = y+z$, to get, ignoring numerical coefficients, $x^{17}(y+z)^0 + x^{16}(y+z)^1 + \cdots + x^0(y+z)^{17}$. Use the binomial theorem again on each of those original terms to see that $(y+z)^k$ expands to $y^k z^0 + y^{k-1}z^1 + \cdots + y^0 z^k$, which has $k+1$ terms. Since all the original terms have distinct powers of x, none of the new terms combine with each other, so there are $1+2+\cdots+18 = \boxed{171}$ terms in total.

35. Since we can rotate the whole bracelet after making it, we may as well pick one charm and fix its position; the other charms can then be placed in any order, for a total of $5! = \boxed{120}$ ways.

36. If the third letter is the kth letter in the alphabet, then the previous two letters can be any distinct two of the preceding $k-1$ letters and must be placed in order, so there are $\binom{k-1}{2}$ possibilities there. The last letter must be one of the $26-k$ succeeding letters, so there are $(26-k)\binom{k-1}{2}$ possibilities with the given third letter. Since N, Y, C, M, and L are the 14th, 25th, 3rd, 13th, and 12th letters in the alphabet, the total number is $12\cdot\binom{13}{2}+1\cdot\binom{24}{2}+23\cdot\binom{2}{2}+13\cdot\binom{12}{2}+14\cdot\binom{11}{2} = 12\cdot 78 + 276 + 23 + 13\cdot 66 + 14\cdot 55 = \boxed{2863}$.

37. Suppose we have k integers; then we can write them as $n+0, n+1, \ldots, n+k-1$, and their sum is $100 = (n+n+\cdots+n) + (0+1+\cdots+(k-1)) = kn + \frac{k(k-1)}{2}$, so we want $200 = k(2n+k-1)$. The two factors on the right differ by $2n-1$, so they have different parities; $200 = 2^3 \cdot 5^2$, so this can only happen with the products $(2^3) \cdot (5^2) = 8\cdot 25$, $(2^3 \cdot 5) \cdot (5) = 40 \cdot 5$, and $(2^3 \cdot 5^2) \cdot (1) = 200 \cdot 1$. Since $n > 1$, $2n-1 > 0$, so the smaller factor is k; thus the first gives $k = 8$ and $n = 9$, the second gives $k = 5$ and $n = 18$, and the last gives $k = 1$ and $n = 100$. Since we want more than one number, we reject $k = 1$, so there are $\boxed{2}$ possibilities.

38. $\overline{AC}$ lies on the line $y = 3$, so the lattice points are $\{(n,3) \mid n \in [3,12] \wedge n \in \mathbb{Z}\}$, which contains $12 - 3 + 1 = 10$ elements. $\overline{AB}$ lies on the line $y = x$, so its lattice points (not counting A) are $\{(n,n) \mid n \in [4,7] \wedge n \in \mathbb{Z}\}$, which contains $7 - 4 + 1 = 4$ elements.

Finally, going from B to C involves moving 5 units along the x-axis and 4 along the y-axis. Since $\gcd(4,5) = 1$, there are no other lattice points along the way. Thus there are $10 + 4 = \boxed{14}$ points in total.

39. Among the integers from 1 to 2006, $\lfloor \frac{2006}{3} \rfloor = 668$ are multiples of 3, $\lfloor \frac{2006}{7} \rfloor = 286$ are multiples of 7, and $\lfloor \frac{2006}{\text{lcm}(3,7)} \rfloor = 95$ are both. Thus $668 + 286 - 95 = 859$ are multiples of 3 or 7, and $2006 - 859 = \boxed{1147}$ are multiples of neither.

40. Suppose $1155 + n^2 = m^2$. Then $m^2 - n^2 = (m-n)(m+n) = 1155$. Assume $m > 0$ (since we don't care about m itself, but only about m^2). All of the factors of 1155 are odd, so any factorization of it into two numbers gives integer values for m and n (respectively, half the sum and half the positive difference of the two factors). Since $1155 = 3^1 \cdot 5^1 \cdot 7^1 \cdot 11^1$, it has $(1+1)(1+1)(1+1)(1+1) = 16$ factors, so there are $\frac{16}{2} = \boxed{8}$ distinct factorizations.

Level 3

41. **Solution 1:** If $z = 0$, we have the following values for (x,y): $(0,0)$, $(0,\pm 1)$, $(\pm 1, 0)$, $(\pm 1, \pm 1)$, $(0, \pm 2)$, $(\pm 2, 0)$, $(\pm 1, \pm 2)$, and $(\pm 2, \pm 1)$, for a total of 21 solutions. If $z = \pm 1$, the same values of (x,y) work, for 42 more solutions. If $z = \pm 2$, then $x^2 + y^2 < 3$, which yields $(0,0)$, $(0, \pm 1)$, $(\pm 1, 0)$, and $(\pm 1, \pm 1)$, for a total of 18 more solutions. Thus there are $\boxed{81}$ solutions in total.

42. Any real number x can be written as $\lfloor x \rfloor + f$, where $f \in [0,1)$. If $f \in [0, \frac{1}{2})$, then $\lfloor 2x \rfloor = 2 \lfloor x \rfloor$; otherwise $\lfloor 2x \rfloor = 2 \lfloor x \rfloor + 1$. Similarly, if $f \in [0, \frac{1}{3})$, $\lfloor 3x \rfloor = 3 \lfloor x \rfloor$; if $f \in [\frac{1}{3}, \frac{2}{3})$, $\lfloor 3x \rfloor = 3 \lfloor x \rfloor + 1$; otherwise $\lfloor 3x \rfloor = 3 \lfloor x \rfloor + 2$. Thus,

$$\lfloor 2x \rfloor + \lfloor 3x \rfloor = 5 \lfloor x \rfloor + \begin{cases} 0, & f \in [0, \frac{1}{3}), \\ 1, & f \in [\frac{1}{3}, \frac{1}{2}), \\ 2, & f \in [\frac{1}{2}, \frac{2}{3}), \\ 3, & f \in [\frac{2}{3}, 1). \end{cases}$$

This tells us that the right side of the equation is not congruent to 4 mod 5. Thus 4 out of every 5 numbers from 1 to 100 work, and there are $\frac{4}{5} \cdot 100 = \boxed{80}$ of those.

43. Each of the numbers is of the form $\overbrace{1\cdots 1}^{\substack{n \\ \text{times}}}\overbrace{5\cdots 5}^{\substack{n-1 \\ \text{times}}}6$. The value of such a number is

$$6 + 5\left(10 + 10^2 + \cdots + 10^{n-1}\right)$$
$$+ \left(10^n + 10^{n+1} + \cdots + 10^{2n-1}\right).$$

Apply the formula for the sum of a finite geometric sequence:

$$= 6 + \frac{5(10^n - 10)}{9} + \frac{10^{2n} - 10^n}{9}$$
$$= \frac{10^{2n} + 4 \cdot 10^n + 4}{9}$$
$$= \left(\frac{10^n + 2}{3}\right)^2$$

so all $\boxed{5}$ are perfect squares.

44. **Solution 1:** Let the lengths of the sides be a, b, and c, with $a \le b \le c$. We know $a + b + c = 15$ and the triangle inequality additionally implies $c < a + b$. Note that $a \le 5$, so let's look at all the cases:

- $a = 1$: $c = 14 - b$ and $1 \le b \le c < b + 1$, so $(b, c) = (7, 7)$.
- $a = 2$: $c = 13 - b$ and $2 \le b \le c < b + 2$, so $(b, c) = (6, 7)$.
- $a = 3$: $c = 12 - b$ and $3 \le b \le c < b + 3$, so $(b, c) = (6, 6)$ or $(5, 7)$.
- $a = 4$: $c = 11 - b$ and $4 \le b \le c < b + 4$, so $(b, c) = (5, 6)$ or $(4, 7)$.
- $a = 5$: $c = 10 - b$ and $5 \le b \le c < b + 5$, so $(b, c) = (5, 5)$.

So there are $\boxed{7}$ triangles.

Solution 2: Let the lengths of the sides be x, y, and z, with $x \le y \le z$. We have $x + y + z = 15$, which defines a plane in 3-space; the positive lattice points within the plane form a triangular grid, depicted in Fig. 6.2. The triangle inequality additionally implies $z < x + y$. The points satisfying all these inequalities form the filled region in Fig. 6.2, and there are $\boxed{7}$ lattice points inside the region or on the solid parts of the border (corresponding to the non-strict inequalities).

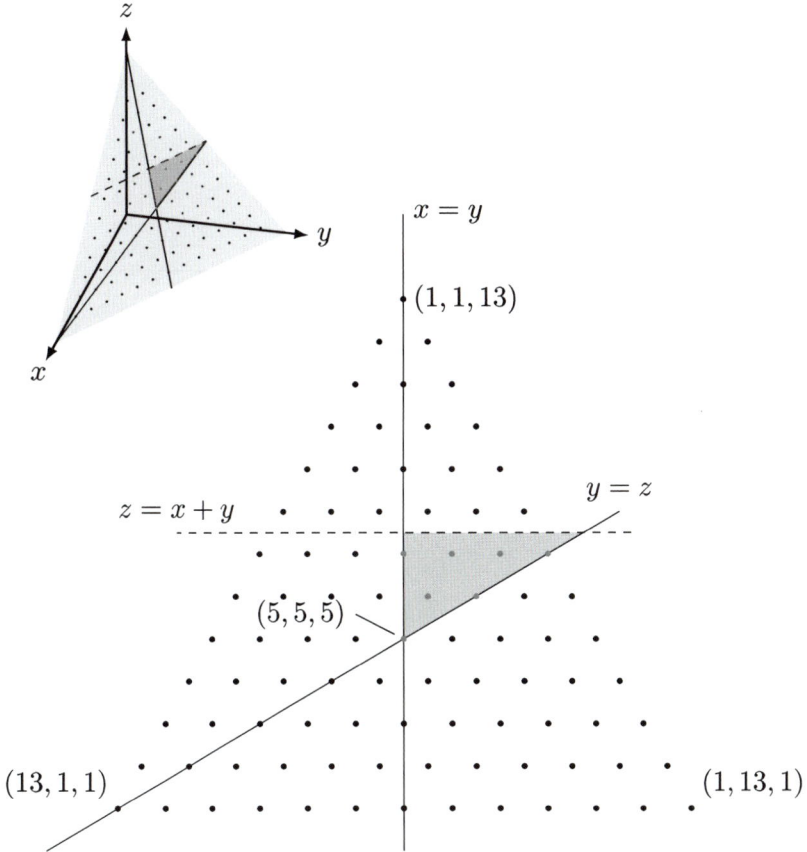

Figure 6.2

Chapter 7

Number Theory

Questions

Level 1

1. Compute the total number of positive integral divisors of 1728.
2. Compute the sum of the three smallest positive integers which have an odd number of divisors.
3. Find the smallest positive multiple of 20 that is a perfect cube.
4. Find all ordered pairs of digits (a, b) such that $\underline{a}\,\underline{9}\,\underline{7}\,\underline{b}$ is divisible by 45.
5. Find the smallest positive multiple of 12 that leaves a remainder of 4 when divided by 52.
6. If the number $\underline{3,1\,6\,5,2\,8\,4,9\,a\,7,6\,3\,4}$ is divisible by 9, find a.
7. The sum of four different prime numbers divides their product. Find the smallest of these prime numbers.

Level 2

8. Find how many ordered pairs of positive integers (x, y) satisfy the equation $3x + 5y = 500$.
9. The equation $19x + 85y = 1985$ is satisfied by the ordered pair of positive integers $(x, y) = (100, 1)$. Find another such pair.
10. Suppose x is a three-digit integer, y is the three-digit integer formed by reversing the digits of x, and $x > y$. Find all values of x such that $x - y$ is a multiple of 5 and the sum of its digits is equal to the sum of the digits of x.

11. Find the smallest whole number which leaves a remainder of 1 when divided by each of the integers between 2 and 10 inclusive.

12. Find how many four-digit positive integers are divisible by 12 and consist of entirely prime digits.

13. Find the smallest positive integer whose cube ends in the digits 03.

14. Find the smallest two-digit natural number whose cube ends in the digits 432.

15. Given that $63^3 = 250047$, compute the smallest integer greater than 63 whose cube ends in the digits 47.

16. Find a four-digit integer whose two rightmost digits are identical, whose two leftmost digits are identical, and which is a perfect square.

17. Find the ordered pair of digits (a, b) such that $\underline{a}\,\underline{9}\,\underline{7}\,\underline{5}\,\underline{b}$ is a multiple of 88.

18. For some natural number n, $n^2 + 2n$ has a units digit of 4. Compute the tens digit of $n^2 + 2n$.

19. Find the largest prime number that is four more than the fourth power of an integer.

20. Compute the sum of all possible values of positive integer x, $x \leq 75$, such that $7^x - 3^x$ is divisible by 10.

21. Compute the remainder when 7^{31} is divided by 4.

22. Find the largest positive integer n for which $\frac{3n+25}{2n-5}$ is a positive integer.

23. Compute the remainder when 2^{150} is divided by 9.

24. Find the ordered pair of digits (a, b) such that $\underline{1,7}\,\underline{1}\,\underline{7,1}\,\underline{7}\,\underline{1,7}\,\underline{a}\,\underline{b}$ is divisible by 99.

25. $k = 1! + 2! + 3! + \cdots + n!$ and k is a perfect square. Compute all possible values of n.

26. Find the four numbers in the set $\{22, 29, 33, 37, 44, 52, 59, 63, 75, 85\}$ whose product is 1,392,754.

27. Olive and William together choose a number between 1900 and 2000 and write each of its positive divisors on ping pong balls. They each then take an equal number of the balls and give the single leftover ball to Chip, who notes that the number on his ball is odd. What are all the possible numbers on his ball?

28. Let n be the smallest positive integer such that removing the leftmost digit of n and placing it after the rightmost digit creates $\frac{n}{4}$. Compute the number of digits in n.

29. If 606, 967, and 1404 are each divided by the positive integer q, where $q > 1$, they all leave the same remainder r. Compute r.

30. There are three consecutive odd positive integers greater than 10 such that the smallest is divisible by 3, the middle by 5, and the largest by 7. Compute the smallest possible value of the smallest of these integers.

31. Find the smallest positive integer c such that $546a + 1365b = 10^6 + c$ for some integers a and b.

Level 3

32. Find the integer n such that $x^2 - x + n$ is a factor of $x^9 + 342x - 6461$.

33. Find the smallest positive integer n that (a) has exactly three distinct prime factors and (b) has exactly eight positive integral factors, which sum to 3696.

34. n consists of the digits from 1 to 6, each used exactly once, such that number formed by the k leftmost digits of n is divisible by k for each k from 1 to 6. Find both possible values of n.

35. How many integers T are there between 1 and 1992 inclusive such that $T^2 - 1$ is divisible by 256?

36. Let $f(m, n) = \frac{7}{2}(m + n) + \frac{3}{2}|m - n|$. Find how many ordered pairs of positive integers (m, n) satisfy $f(m, n) = 1992$.

Answers

Level 1

1. If the prime factorization of a number is $p_1^{k_1} p_2^{k_2} p_3^{k_3} \cdots$, the total number of divisors it has is $(k_1 + 1)(k_2 + 1)(k_3 + 1) \cdots$. Since $1728 = 12^3 = 2^6 3^3$, it has $7 \cdot 4 = \boxed{28}$ divisors.

2. An integer has an odd number of divisors if and only if it is a perfect square. The sum of the smallest three perfect squares is $1 + 4 + 9 = \boxed{14}$.

3. In the prime factorization of a perfect cube, all the exponents must be multiples of 3. Since $20 = 2^2 5^1$, the smallest perfect cube it divides is $2^3 5^3 = \boxed{1000}$.

4. Since the number is divisible by 5, $b = 0$ or 5. Since the number is divisible by 9, the sum of its digits must be divisible by 9. If $b = 0$, the digit sum is $a + 16$, so $a = 2$; if $b = 5$, the digit sum is $a + 21$, so $a = 6$. Thus $(a, b) = \boxed{(2, 0) \text{ or } (6, 5)}$.

5. We need to find positive a and non-negative b (both integers) such that $12a = 52b + 4$, or, equivalently, $3a = 13b + 1$. So $13b \equiv b \equiv -1$ (mod 3), so $b = 2$, which leads to $a = \frac{13 \cdot 2 + 1}{3} = 9$, and $12a = \boxed{108}$.

6. The number is divisible by 9 if and only if the sum of its digits is; the sum is $58 + a$, so we want $a = \boxed{5}$. (The arithmetic can be simplified by "casting out nines," i.e., removing 9s and groups of digits which sum to 9. Here, we can drop every digit but one of the 4s.)

7. If all the primes were odd, then their sum would be even and their product would be odd, and it would be impossible for their sum to divide their product. One of the primes must therefore be even, which means it must be $\boxed{2}$, the smallest prime, and therefore the smallest of these primes. (One set of prime numbers having this property is $\{2, 3, 11, 17\}$.)

Level 2

8. **Solution 1:** Rearranging gives $y = 100 - \frac{3x}{5}$. Thus x must be a multiple of 5; it must be positive and, in order for y to be positive, strictly less than $\frac{100 \cdot 5}{3}$, so it can take on $\lceil \frac{500}{5} \rceil - 1 = \boxed{33}$ different values.

   Solution 2: One solution (ignoring the requirement for positive numbers) is $(x, y) = (0, 100)$. From any solution we can derive another solution by increasing x by 5 and decreasing y by 3, since $3(x + 5) + 5(y - 3) = 3x + 5y$. Doing this once immediately leads to a solution with positive integers, and we can do it $\lceil \frac{100}{3} \rceil - 1 = \boxed{33}$ times in total before y stops being positive.

9. **Solution 1:** As in the previous solution, we can take any solution, decrease x by 85, increase y by 19, and end up with another solution; doing that here gives $\boxed{(15, 20)}$.

   Solution 2: Taking the original equation modulo 19 gives $85y \equiv 1985$, so $9y \equiv 9$ and $y \equiv 1$. Thus the possible values of y are $1, 20, \ldots$, and $y = 20$ leads to the solution $\boxed{(15, 20)}$.

10. Let $x = a\,b\,c$, so $y = c\,b\,a$. We have $a > c > 0$, since both numbers have three digits and $x > y$. Since $x - y = 99(a - c)$ is a multiple of 5, so is $a - c$, and we must have $a - c = 5$. Thus $x - y = 495$ and the sum of the digits of x is $4 + 9 + 5 = 18$. Given $a - c = 5$, (a, c) can be any of $(9, 4)$, $(8, 3)$, $(7, 2)$, or $(6, 1)$, but if it is $(6, 1)$, then no value of b can make $a + b + c = 18$. Otherwise, there is a valid value of b, leading to the numbers $\boxed{954, 873, \text{ and } 792}$.

11. We have that $n - 1$ is divisible by all of the numbers from 2 to 10, so the smallest possible value of n is

$$1 + \text{lcm}(2, 3, \ldots, 10) = 1 + \text{lcm}(2, 3, 2^2, 5, 2 \cdot 3, 7, 2^3, 3^2, 2 \cdot 5)$$
$$= 1 + 2^3 \cdot 3^2 \cdot 5 \cdot 7$$
$$= \boxed{2521}.$$

12. The allowable digits are 2, 3, 5, and 7. An integer is a multiple of 12 if and only if it is a multiple of both 3 and 4. It is a multiple of 4 if and only if the number formed by its two rightmost digits is; this means its units digit must be 2 and its tens digit can be 3, 5, or 7. Meanwhile, it is a multiple of 3 if and only if the sum of its digits is a multiple of 3. Call the first 2 digits a, b, so if the last two digits are 32, $a + b \equiv 1 \bmod 3$, giving 2232, 2532, 5232, 3732, 7332, 5532. If the last two digits are 52, $a + b \equiv 2 \bmod 3$, giving 2352, 3252, 3552, 5352, 7752. If the last two digits are 72, $a + b \equiv 0 \bmod 3$, giving 3372, 2772, 7272, 5772, 7572 for $\boxed{16}$ possibilities.

13. Write $n = 10t + u$, where t is a non-negative integer and u is a digit. Then $n^3 = 1000t^3 + 300t^2u + 30tu^2 + u^3$. Since the last digit must be 3, we can look at this modulo 10 to see $u^3 \equiv 3 \pmod{10}$, which is only satisfied by $u = 7$. Then, modulo 100,

$$3 \equiv n^3 \equiv 30tu^2 + u^3 \equiv 70t + 43 \quad \bmod 100$$
$$60 \equiv 70t \quad \bmod 100$$
$$6 \equiv 7t \quad \bmod 10,$$

so $t \equiv 8 \pmod{10}$, which leads to the smallest non-negative solution of $t = 8$, and $n = \boxed{87}$.

14. Let the number we want be $n = \underline{t}\,\underline{u} = 10t + u$. Then $n^3 = 1000t^3 + 300t^2u + 30tu^2 + u^3$. Since the last digit must be 2, we can look at this modulo 10 to see $u^3 \equiv 2 \pmod{10}$, which is only satisfied by $u = 8$. Then

$$32 \equiv n^3 \equiv 30tu^2 + u^3 \equiv 20t + 12 \quad \bmod 100$$
$$20 \equiv 20t \quad \bmod 100$$
$$1 \equiv t \quad \bmod 5,$$

so, since t is a digit, $t = 1$ or 6 and $n = 18$ or 68. Computing the cubes shows that 18^3 has a different hundreds digit, but $n = \boxed{68}$ works.

15. **Solution 1:** Since only a number ending in 3 can have a cube ending in 7, we can write the number we want as $n = 10a + 3$

for some integer a. Then $n^3 = 1000a^3 + 900a^2 + 270a + 27$; the first two terms do not affect the last two digits of the result, so we need $270a + 27 = 100b + 47$ for some integer b, or $27a = 10b + 2$. So we need a multiple of 27 (or, equivalently, 7) with a units digit of 2. This is satisfied by $a = 6, 16, 26, \ldots$, leading to $n = 63$ and then $\boxed{163}$.

Solution 2: Since only a number ending in 3 can have a cube ending in 7, we can write the number we want as $n = 10k + 63$ for some positive k. We want n^3 to end in the same two digits as 63^3, so $n^3 \equiv 63^3 \pmod{100}$. Since $a^3 - b^3 = (a - b)(a^2 + ab + b^2)$ in general and $n - 63 = 10k$, we have

$$n^3 - 63^3 \equiv 0 \quad \mathrm{mod}\ 100$$
$$10k \cdot (n^2 + 63n + 63^2) \equiv 0 \quad \mathrm{mod}\ 100$$
$$k \cdot (n^2 + 3n + 3^2) \equiv 0 \quad \mathrm{mod}\ 10.$$

Since $n \equiv 3 \pmod{10}$, we have $n^2 + 3n + 3^2 \equiv 3^2 + 3 \cdot 3 + 3^2 \equiv 7$, so

$$7k \equiv 0 \quad \mathrm{mod}\ 10.$$

Since 7 and 10 are relatively prime, this means $k \equiv 0 \pmod{10}$, so the smallest positive solution is $k = 10$ and $n = \boxed{163}$.

16. Write the number as $n^2 = \underline{a}\,\underline{a}\,\underline{b}\,\underline{b} = 1100a + 11b = 11 \cdot \underline{a}\,\underline{0}\,\underline{b}$. Thus n^2 is divisible by 11, so n must be as well, and so n^2 is in fact divisible by 11^2. Thus $\underline{a}\,\underline{0}\,\underline{b} = \frac{n^2}{11} = 11 \left(\frac{n}{11}\right)^2$ is a perfect square multiplied by 11, so the divisibility test for 11 gives $a + b = 11$. We have,

$$\underline{a}\,\underline{0}\,\underline{b} = \underline{11 - b}\,\underline{0}\,\underline{b}$$
$$= \underline{10 - b}\,\underline{10 - b}\,\underline{0} + \underline{b}\,\underline{b}$$
$$= 11 \cdot \underline{10 - b}\,\underline{b}.$$

(For example, $506 = 440 + 66 = 11 \cdot 46$.) All the "digits" in the computation are actually valid digits (i.e., between 0 and 9) because $a + b = 11$ implies $a \geq 2$ and $b \geq 2$. Thus $\underline{10 - b}\,\underline{b}$ is a two-digit perfect square whose digits sum to 10. The only such number is 64, so $b = 4$, $a = 7$, and $n = \boxed{7744}$.

17. The number is divisible by 8 and 11. Thus the number formed by the rightmost three digits is divisible by 8; since $75 \equiv 3 \pmod 8$, $750 \equiv 3 \cdot 10 \equiv 6 \pmod 8$, so $b = 2$. The divisibility test for 11 gives that $a + 7 + 2 - (9 + 5) = a - 5$ is divisible by 11, so $a = 5$ and $(a, b) = \boxed{(5, 2)}$.

18. Since $n^2 + 2n$ ends in 4, $n^2 + 2n + 1 = (n+1)^2$ ends in 5, so $n+1$ ends in 5. The square of a number ending in 5 always ends in 25, so the tens digit is $\boxed{2}$.

19. For any n,
$$n^4 + 4 = n^4 + 4n^2 + 4 - 4n^2$$
$$= (n^2 + 2)^2 - (2n)^2$$
$$= (n^2 + 2n + 2)(n^2 - 2n + 2).$$

   For this to be prime, one of these factors must be 1. If it is the first, then $n = -1$; if it is the second, then $n = 1$; either way, the prime is $1^4 + 4 = \boxed{5}$.

20. The units digits of successive powers of 7 are $7, 9, 3, 1 \ldots$ and for 3 are $3, 9, 7, 1 \ldots$. Since the second and fourth digits (etc.) of each are the same. $7^x - 3^x$ is divisible by 10 when $x \equiv 0 \bmod 4$ or $x \equiv 2 \bmod 4$. So $x = 0, 2, 4, 6, \ldots, 74$. We can write their sum as $2 + 4 + 6 + \cdots + 74 = 2 \cdot (1 + 2 + 3 + \cdots + 37)$. The formula for the sum of the first n positive integers is $\frac{n(n+1)}{2}$, so we get $2 \cdot \frac{37 \cdot 38}{2}$ or $37 \cdot 38 = \boxed{1406}$.

21. Modulo 4, we have $7 \equiv -1$ and so $7^{31} \equiv (-1)^{31} \equiv -1 \equiv \boxed{3}$.

22. Double the given quantity to see that $\frac{6n+50}{2n-5} = \frac{(6n-15)+65}{2n-5} = 3 + \frac{65}{2n-5}$ is an even integer. Thus $2n-5$, which increases with n, divides 65 with an odd quotient; the largest it can be is 65 itself, leading to $n = \frac{65+5}{2} = \boxed{35}$.

23. Modulo 9, we have $2^{150} \equiv \left(2^3\right)^{50} \equiv (-1)^{50} \equiv \boxed{1}$.

24. **Solution 1:** We have $100 \equiv 1 \pmod{99}$, and so, modulo 99,
$$0 \equiv 17 \cdot 100^4 + 17 \cdot 100^3 + 17 \cdot 100^2 + 17 \cdot 100 + \underline{a}\,\underline{b}$$
$$\equiv 17 + 17 + 17 + 17 + \underline{a}\,\underline{b}$$
$$\equiv 68 + \underline{a}\,\underline{b}.$$

   So we want $\underline{a}\,\underline{b} \equiv 31 \pmod{99}$, which is satisfied by $\underline{a}\,\underline{b} = 31$, so $(a, b) = \boxed{(3, 1)}$.

   Solution 2: Since the number is a multiple of 9, the sum of its digits, which is $32 + a + b$, is a multiple of 9, so $a + b$ must be 4 or 13. Since it is a multiple of 11, the difference of alternating digits, which is $a - b - 24$, is a multiple of 11, so $a - b = 2$ or -9. Since a is the average of $a + b$ and $a - b$ and b is half the difference between them, the two must have the same parity and differ by at most 18, this leaves only $a + b = 4$ and $a - b = 2$, so $(a, b) = \boxed{(3, 1)}$.

25. Anything from 5! on will have a 0 as the units digit. So only $1! + 2! + 3! + 4!$ will contribute to the units digit, and $1 + 2 + 6 + 24 = 33$, with a units digit of 3. No perfect square has a units digit of 3, so we only have to look at $1! = 1, 1! + 2! = 3$, and $1! + 2! + 3! = 9$. So n can be $\boxed{1}$ or $\boxed{3}$.

26. The digit sum of the number is not divisible by 3, so neither is the number itself, which eliminates 33, 63, and 75. Since the last two digits of the number are 54, it is a multiple of neither 4 nor 5, which further eliminates 44, 52, and 85. The remaining four numbers are $\boxed{22, 29, 37, \text{ and } 59}$.

27. The number of divisors of the number is odd, so the number must be a perfect square; the only one in the given range is $1936 = 44^2 = 2^4 \cdot 11^2$. Its odd divisors are $11^0, 11^1$ and 11^2, or $\boxed{1, 11, \text{ and } 121}$.

28. Suppose that n has d digits, and its leftmost digit is a. Then adding a new a to the right turns n into $10n + a$, and removing the old a from the left turns that into $10n + a - 10^d a = 10n - (10^d - 1)a$ which equals $\frac{n}{4}$, so $39n = 4(10^d - 1)a$. So we need $4(10^d - 1)a$ to be a multiple of 39, i.e., of both 3 and 13. Neither 4 nor a can be a multiple of 13; since $10^d - 1$ is always a multiple of 3, we just need to find the smallest d for which it is a multiple of 13. Modulo 13, we have $10^1 \equiv -3, 10^2 \equiv 9, 10^3 \equiv -27 \equiv -1$, and $10^6 \equiv 1$, so $d = \boxed{6}$. (Having checked 10^6, we don't have to check 10^4 and 10^5 because if $10^a \equiv 10^b$, then $10^{a-b} \equiv 1$, and we already checked 10^1 and 10^2.)

29. Since all three numbers are congruent modulo q, the difference between any two of them is a multiple of q. We have $967 - 606 = 361 = 19^2$ and $1404 - 967 = 437 = 21^2 - 4 = 19 \cdot 23$. The only common factor of these differences (besides 1) is 19, so $q = 19$ and we compute, e.g., $967 - 19 \cdot 50 = \boxed{17}$.

30. If the smallest of the integers is a, the other two are $a + 2$ and $a + 4$. Since a is a multiple of 3, so is $a - 3$; since $a + 2$ is a multiple of 5, so is $a + 2 - 5 = a - 3$; since $a + 4$ is a multiple of 7, so is $a + 4 - 7 = a - 3$. Thus $a - 3$ is an even (since a is odd) multiple of $\text{lcm}(3, 5, 7) = 105$. Thus $a - 3 = 210$ and $a = \boxed{213}$.

31. Factor the given coefficients to get $273(2a + 5b)$; this can equal any multiple of $273 \cdot \gcd(2, 5) = 273$. We have

$$273 = 3 \cdot 7 \cdot 13$$

$$10^6 - 1 = (10^3 + 1)(10^3 - 1) = (7 \cdot 11 \cdot 13) \cdot (3 \cdot 333),$$

so $10^6 \equiv 1 \pmod{273}$. Thus the least multiple of 273 which is greater than 10^6 is $10^6 + 272$, so $c = \boxed{272}$.

Level 3

32. If $f(x)$ is a factor of $g(x)$, then, for any integer m, $|f(m)|$ is a factor of $|g(m)|$. Substituting $x = 0$ and $x = 1$ gives that n divides both 6461 and $6461 - 342 - 1$, so it divides their difference, which is $343 = 7^3$. Since 49 does not divide 6461, we have $n = \pm 1$ or ± 7. Substituting $x = -1$ gives that $|n + 2|$ divides 6804, which eliminates -7; substituting $x = -2$ gives that $n + 6$ divides 7657, which eliminates ± 1. Thus $n = \boxed{7}$.

33. Write $n = p^a q^b r^c$, where p, q, and r are the three prime factors of n. The number of divisors of n is $(a+1)(b+1)(c+1) = 8$; since a, b, and c are all at least 1, we must have $a = b = c = 1$. The sum of the divisors of n is (see the appendix)

$$\frac{p^2 - 1}{p - 1} \cdot \frac{q^2 - 1}{q - 1} \cdot \frac{r^2 - 1}{r - 1} = (p+1)(q+1)(r+1) = 3696 = 2^4 \cdot 3 \cdot 7 \cdot 11.$$

The easiest thing to try, $p = 2$ and $q = 3$, leads to $r = 307$ (a prime) and so $n = \boxed{1842}$; checking the other possibilities shows that that is in fact the smallest possible result.

A sketch of why you might believe this without actually checking everything: given a fixed value of $(p + 1)(q + 1)(r + 1)$, we want to minimize pqr, or, equivalently, $\frac{p}{p+1} \cdot \frac{q}{q+1} \cdot \frac{r}{r+1}$. For real x, $\frac{x}{x+1}$ goes to 0 as x goes to 0, but it only approaches 1 even as x gets arbitrarily large. So perhaps it makes sense that, since our variables are constrained to be integers, we can find the minimum by making two of the variables as small as possible and letting the other be large.

34. We can deduce the following statements.

- The 5th digit is 5.
- The 2nd, 4th, and 6th digits are even, so the 1st and 3rd are odd and thus are 1 and 3 in some order.
- The 2nd digit is 2, since neither $1 + 3 + 4$ nor $1 + 3 + 6$ is divisible by 3.
- The 4th digit is 6, since neither 14 nor 34 is divisible by 4.
- The 6th digit is 4.

The 1 and 3 may come in either order, leaving $\boxed{123654 \text{ and } 321654}$.

35. We have $(T-1)(T+1) = 2^8 m$ for some positive integer m. So $T-1$ and $T+1$ must be even, but only one of them is divisible by 4; thus that one must in fact be divisible by $2^7 = 128$, and it is sufficient for that to be the case. Thus $T = 0 \cdot 128 + 1$ works, and we have $15 \cdot 128 < 1992 < 16 \cdot 128$, so, for $k = 1, \ldots, 15$, both $T = 128k + 1$ and $T = 128k - 1$ work. Thus there are $15 \cdot 2 + 1 = \boxed{31}$ possibilities.

36. **Solution 1:** Since $f(m, m) = 7m$ and 7 does not divide 1992, there are no solutions with $m = n$. Since $f(m, n) = f(n, m)$, we may assume $m > n$ and double the number of solutions at the end. We can write $m = n + k$. Thus we want $k > 0$, $n > 0$, and

$$f(m, n) = \frac{7}{2}(m + n) + \frac{3}{2}(m - n) = 5m + 2n = 5k + 7n = 1992.$$

The solution with minimal n is $(k, n) = \left(\frac{1985}{5}, 1\right) = (397, 1)$. Taking any solution and increasing n by 5 while decreasing k by 7 gives a new solution; we can do this $\left\lceil \frac{397}{7} \right\rceil - 1 = 56$ times before k is no longer positive, for a total of 57 solutions. Finally, we apply the aforementioned doubling to get $57 \cdot 2 = \boxed{114}$ solutions.

Solution 2: As before, we assume $m > n$ and so $1992 = 5m + 2n$. Modulo 5, this becomes $2n \equiv 2$ and so $n \equiv 1$; thus we can write $n = 5t + 1$ for some integer t. Substituting this back in gives $1992 = 5m + 10t + 2$, or $m = 398 - 2t$. Now, we must have $n = 5t + 1 \geq 1$, which implies $t \geq 0$, and $m = 398 - 2t \geq 1$, which implies $t \leq 198$. We also need $m > n$, or $398 - 2t > 5t + 1$, so $397 > 7t$, which implies $t \leq 56$. Thus we have a valid solution for each $0 \leq t \leq 56$, so there are 57; apply the doubling to get $57 \cdot 2 = \boxed{114}$ solutions.

Chapter 8

Probability

Questions

Level 1

1. Two tiles marked with the letter T and two with the letter O are placed in an urn, and the urn is shaken. All the tiles are withdrawn and placed in order on a rack. Compute the probability that the two Os are not adjacent and the two Ts are not adjacent.

2. Two fair six-sided dice are thrown. Compute the probability that the product of the numbers showing on top is 12.

3. Two fair six-sided dice are thrown. Compute the probability that the product of the numbers showing on top is not a prime.

4. The numbers 3, 4, 5, 6, 7, 8, and 9 are written on individual cards and are placed in a hat. If four cards are drawn without replacement, compute the probability that the product of the numbers on the cards chosen is odd.

5. If the probability that it will not rain is the square of the probability that it will rain, compute the probability that it will rain.

6. A commuter train runs along a straight line path at constant speed. It starts at town A and passes through at towns B, C, D, and E before arriving at city F, where it immediately reverses direction and goes back toward A, and then back to F again, back and forth all day. If the distances between any two adjacent stops are equal, compute the probability that, at a random point in time, the train either will arrive at F next or left from F most recently.

7. A fair die is rolled 6 times. Compute the probability of rolling a number less than 3 exactly 5 times.

8. If the digits a and b are chosen uniformly at random, compute the probability that the integer $\underline{1}\,\underline{a}\,\underline{7}\,\underline{b}$ is divisible by 4.

9. Raoul tosses 2 fair coins. He shows you one of the coins, and it came up heads. Compute the probability that the other coin came up tails.

10. Mrs. Daaé has three children. One day, you see her on the street with a little girl. She introduces you and tells you that the girl is her middle child. Compute the probability that the other two children are girls, assuming that the probability of a child being born a girl is $\frac{1}{2}$.

Level 2

11. Ten cards, numbered $1, 2, 3, \ldots, 10$, are placed face down on a table. One card is drawn at random, its number recorded, and then replaced face down. A card is drawn again. Compute the probability that the number on the second draw is greater than the number on the first draw.

12. On a one question quiz, the probability of Audrey getting the right answer is $\frac{1}{4}$. The probability of Seymour getting the right answer is $\frac{1}{3}$. The probability of Orin getting the right answer is $\frac{1}{2}$. Compute the probability that exactly one of them gets the right answer on the quiz.

13. Mr. Cervantes is a teacher. To determine one student's grade, he rolls two ten-sided dice with the digits from 0 to 9 on the sides. He uses the higher roll for the tens digit and the lower roll for the units digit. (If the rolls are equal, then both digits are that roll.) Compute the probability that the resulting score is at least 65.

14. Rona guesses on every question of a five-question true–false quiz. Find the probability that she gets at least 60% correct.

15. Three distinct integers are chosen randomly from the set $\{1, 2, 3, \ldots, 10\}$. Compute the probability that their sum is a multiple of 3.

16. Consider two very poor machines. One has three parts, and will work if and only if at least two of the parts do not fail. One part fails with probability $\frac{3}{5}$ and each of the other two fails with probability p. The second machine fails with probability $p^2 - p + 1$. If the probability of the machines working is the same, find p.

17. Suppose that two dice are "loaded" in such a way that the probability of rolling a face with k dots is proportional to k. Compute the probability of rolling a sum of seven on the two dice.

18. Five pennies, five nickels, and five dimes are in a box. Three coins are drawn at random without replacement. Find the probability that the total value of these coins is less than 15 cents.

19. After two fair six-sided dice are tossed, a total of 10 faces of the dice are visible. Compute the probability that the sum of all spots that can be seen on those ten faces is divisible by 7.

20. A circle of radius $\frac{1}{2}$ is randomly dropped onto the Cartesian plane. Find the probability that there is a lattice point inside the circle.

21. Statements A and B are independent. The probability that A is true is $\frac{\sqrt{3}}{4}$. The probability that B is true is $\frac{1}{\sqrt{3}}$. Compute the probability that $A \to B$ is true.

22. Of all students taking a test that has a passing grade of 65, those who passed had an average grade of 74, while those who failed had an average grade of 58. If the average of all students was 70, compute the probability that a randomly chosen student in this class passed the test.

23. In the annual Elfland Lottery, a natural number is picked at random, with the probability of picking the number n given by $\frac{1}{2^n}$. Compute the probability that the number picked is even but not divisible by 5.

24. Let $N = 3^a + 7^b$. If a and b are integers, not necessarily different, chosen uniformly at random from 1 to 100 inclusive, compute the probability that the units digit of N is 4.

25. Benjamin, Lucy, and Johanna, in that order, take turns rolling a pair of dice. The first one to roll a total of 9 on the two dice is the winner. The game continues until someone wins. Compute the probability that Lucy wins.

26. Elle and Warner are going to play a game. They put cards with the letters H, A, R, V, A, R, and D in a hat (each of the seven cards has one letter on it and there are two cards with A's or R's). They will alternate picking cards randomly, one at a time, without replacement, until someone wins by picking an A. If Elle picks first, compute the probability that she wins.

27. Helen rolls one fair six-sided die. Jim rolls two fair six-sided dice. Compute the probability that the sum of the numbers showing on Jim's dice equals the number showing on Helen's die.

28. The planet Melmac has a year that is but 8 days long. If three residents of Melmac are in a room, compute the probability that at least two were born on the same day of the year.

29. If an integer is chosen uniformly at random between 100 and 1000 exclusive, find the probability that is divisible by 2, 3, and 11, but not by any perfect cubes greater than 1.

30. A single fair six-sided die is thrown until a 6 appears. Compute the probability that an even number of throws is needed.

Level 3

31. Compute the probability of getting exactly three heads in six tosses of a fair coin, if we know for sure that at least one head will appear in the first three tosses.

32. A coin is flipped until two tails have occurred. If the first tail occurred on the ath flip and the second on the bth, compute the probability that $b = 4a$.

33. Felicia has a fair six-sided die. The numbers $1, 2, 3,$ and 4 appear on four of the faces, while the other two are blank. Nathan has a fair six-sided die showing the numbers 1 through 6. Patrick has a fair 12-sided die showing the numbers 1 through 12. Felicia, Nathan, and Patrick take turns rolling their dice in that order until someone wins by rolling a 1. (If Felicia's die comes up blank, she keeps rolling until it shows a number.) Compute the probability that Patrick wins.

34. Chris, John, and Kim sit at a circular table with 14 other people. If everyone sits down in random seats, compute the probability that Chris is next to Kim or John.

35. A coin is weighted in such a way that the probability of getting heads on its nth flip is $\frac{1}{2^n}$. If the probability of getting 9 heads and 1 tail in 10 flips is $\frac{a}{2^{55}}$, compute a.

Answers

Level 1

1. There are $\binom{4}{2} = 6$ possible arrangements, all equally likely: $TTOO$, $TOTO$, $TOOT$, $OTTO$, $OTOT$, and $OOTT$. Only $TOTO$ and $OTOT$ satisfy the condition, so the probability is $\frac{2}{6} = \boxed{\frac{1}{3}}$.

2. The possible pairs of numbers that multiply to 12 are $(2,6)$, $(3,4)$, $(4,3)$, and $(6,2)$, out of $6 \cdot 6 = 36$ possible outcomes. Thus the probability is $\frac{4}{36} = \boxed{\frac{1}{9}}$.

3. The product is a prime if and only if one number is 1 and the other is 2, 3, or 5. This can happen in 6 different ways, so the probability that it does not is $1 - \frac{6}{36} = \boxed{\frac{5}{6}}$.

4. **Solution 1:** The product is odd if and only if all the numbers chosen are odd. The probability that the first is odd is $\frac{4}{7}$; if it is, then 3 out of 6 left in the hat are odd, so the probability that the second number is odd is $\frac{3}{6}$, and so on, so the overall probability is $\frac{4}{7}\frac{3}{6}\frac{2}{5}\frac{1}{4} = \boxed{\frac{1}{35}}$.

   Solution 2: Out of all 4-card subsets of the 7 cards, of which there are $\binom{7}{4} = 35$, only the one consisting of exactly the four odd numbers has an odd product. Thus the probability is $\boxed{\frac{1}{35}}$.

5. Let p be the probability it will rain, so p^2 is the probability that it will not rain. Since it must either rain or not, $p^2 + p = 1$, or $p^2 + p - 1 = 0$, which has positive root $\boxed{\frac{-1+\sqrt{5}}{2}}$.

6. The train satisfies the condition any time it is on the portion of the track between E and F, which takes up $\frac{1}{5}$ of the length of the whole track. Since the train always runs at the same speed and always goes all the way from one end of the track to the other, it spends $\boxed{\frac{1}{5}}$ of its time in this portion.

7. The probability of rolling less than a 3 on any one roll is $\frac{1}{3}$, so the probability of doing so exactly 5 times out of 6 is $\binom{6}{5} \left(\frac{1}{3}\right)^5 \left(\frac{2}{3}\right) = \boxed{\frac{4}{243}}$.

8. Since a and b can each be chosen in ten ways, there are 100 numbers altogether. By the divisibility rule for 4, b must be 2 or 6. Since there are 10 choices for a, we have 20 choices for these numbers. So the probability is $\frac{20}{100} = \boxed{\frac{1}{5}}$.

9. The possible outcomes of the two coins are HH, HT, TH, and TT. Since one of the coins came up H, the outcome TT is impossible. Of the three outcomes remaining, two have a T besides the H, so the probability is $\boxed{\frac{2}{3}}$. (It might seem like the answer ought to be $\frac{1}{2}$,

since each coin comes up tails with probability $\frac{1}{2}$, but that does not happen because, as far as we know, the coins are indistinguishable, so we do not know whether the outcome was HT or TH.)

10. Since the probability of each child being a girl is $\frac{1}{2}$, the probability of both the first and third children being girls is $\frac{1}{2} \cdot \frac{1}{2} = \boxed{\frac{1}{4}}$.

Level 2

11. There are 100 possible outcomes for the values of the two cards. Of these, there are 9 successes if 1 is drawn first, 8 if 2 is drawn first, and so on, down to 0 if 9 is drawn first. Thus the probability is $\frac{9+8+\cdots+0}{100} = \frac{45}{100} = \boxed{\frac{9}{20}}$.

12. We want the probability that Audrey gets the right answer but the other two do not, or Seymour gets the right answer but the other two don't, or Orin gets the right answer but the other two don't. This is

$$\frac{1}{4}\left(1-\frac{1}{3}\right)\left(1-\frac{1}{2}\right) + \frac{1}{3}\left(1-\frac{1}{4}\right)\left(1-\frac{1}{2}\right)$$

$$+ \frac{1}{2}\left(1-\frac{1}{4}\right)\left(1-\frac{1}{3}\right)$$

$$= \frac{1}{4}\cdot\frac{2}{3}\cdot\frac{1}{2} + \frac{1}{3}\cdot\frac{3}{4}\cdot\frac{1}{2} + \frac{1}{2}\cdot\frac{3}{4}\cdot\frac{2}{3}$$

$$= \boxed{\frac{11}{24}}.$$

13. We can compute the probability of scoring a 66 or less, subtract the probability of scoring 66 or 65, and subtract the result from 1. To score a 66 or less, both rolls must be 6 or less, which occurs with probability $\frac{7}{10} \cdot \frac{7}{10} = \frac{49}{100}$. To get exactly 66, both rolls must be 6, with occurs with probability $\frac{1}{10} \cdot \frac{1}{10} = \frac{1}{100}$. To get exactly 65, one roll must be 6 and the other must be 5, which occurs with probability $\frac{2}{100}$. So the probability of scoring 65 or above is $1 - \frac{49-1-2}{100} = \frac{54}{100} = \boxed{\frac{27}{50}}$.

14. **Solution 1:** The probability of getting exactly k questions right is $\binom{5}{k}\left(\frac{1}{2}\right)^5 = \frac{1}{32}\binom{5}{k}$, so the probability of getting at least 60%, or 3, questions right is $\frac{\binom{5}{5}+\binom{5}{4}+\binom{5}{3}}{32} = \frac{1+5+10}{32} = \boxed{\frac{1}{2}}$.

Solution 2: We want 3, 4, or 5 correct and we do not want 2, 1, or 0 correct; by symmetry (i.e., the probability of getting k correct equals the probability of getting $(5-k)$ correct), the probability is $\boxed{\frac{1}{2}}$.

15. **Solution 1:** Among the integers from 1 to 10, there are 3 congruent to 0 modulo 3, 4 congruent to 1, and 3 congruent to 2. The sum of three integers is a multiple of 3 if they are either all the same or all different modulo 3. There are $\binom{3}{3} + \binom{4}{3} + \binom{3}{3} = 6$ ways to choose three that are all the same and $3 \cdot 4 \cdot 3 = 36$ ways to choose three that are all different, and there are $\binom{10}{3} = 120$ ways to choose any three at all. Thus the probability is $\frac{6+36}{120} = \boxed{\frac{7}{20}}$.

Solution 2: We can simply count the possibilities. The lowest possible sum is $1+2+3 = 6$ and the largest is $8+9+10 = 27$, so we only need to consider the multiples of 3 between 6 and 27. In fact, we can take any set of three integers whose sum is s and replace each element n with $11-n$ to get a new set whose sum is $33-s$, so we only need to consider sums of 6, 9, 12, and 15. By careful enumeration, you can find that these are achievable with 1, 3, 7, and 10 sets respectively, so the probability is $\frac{2\cdot(1+3+7+10)}{\binom{10}{3}} = \frac{42}{120} = \boxed{\frac{7}{20}}$.

16. The second machine works with probability $1-(p^2-p+1) = (1-p)p$. The first machine will work if all parts work or if any combination of two work and the other fails. This is equal to the probability of the second machine working, so

$$(1-p)p = \frac{2}{5}(1-p)(1-p) + \frac{3}{5}(1-p)(1-p) + \frac{2}{5}(1-p)p$$
$$+ \frac{2}{5}p(1-p)$$
$$= (1-p)\frac{2-2p+3-3p+2p+2p}{5}$$
$$= (1-p)\frac{5-p}{5},$$

which is satisfied if $1-p=0$ or $p = \frac{5-p}{5}$, so $p = \boxed{1 \text{ or } \frac{5}{6}}$.

17. Suppose p_k is the probability of rolling the face with k dots. Then $p_k = ck$ for some c and each k; since $\sum p_k = 1$, we have $1 = c + 2c + \cdots + 6c = 21c$, so $p_k = \frac{k}{21}$. We can get a sum of 7 by rolling 1 and 6 in either order, with probability $2 \cdot \frac{1}{21} \cdot \frac{6}{21}$, rolling 2 and 5, with probability $2 \cdot \frac{2}{21} \cdot \frac{5}{21}$, or rolling 3 and 4, with probability $2 \cdot \frac{3}{21} \cdot \frac{4}{21}$. Thus, the total probability is $\frac{2}{21^2}(6 + 10 + 12) = \boxed{\frac{8}{63}}$.

18. The total number of ways to select three coins is $\binom{15}{3} = 455$. For their total value to be less than 15 cents, we can draw

   - 2 pennies and 1 dime: $\binom{5}{2}\binom{5}{1} = 50$ ways,
   - 1 penny and 2 nickels: $\binom{5}{1}\binom{5}{2} = 50$ ways,
   - 2 pennies and 1 nickel: $\binom{5}{2}\binom{5}{1} = 50$ ways,
   - 3 pennies: $\binom{5}{3} = 10$ ways,

   for a total probability of $\frac{50+50+50+10}{455} = \frac{160}{455} = \boxed{\frac{32}{91}}$.

19. The sum of all the spots on the two dice is $2 \cdot (1 + \cdots + 6) = 42$, which is divisible by 7, so the sum of the visible faces is divisible by 7 if and only if the sum of the two non-visible faces on the bottom is. The only possible multiple of 7 that two faces can sum to is 7 itself; if we want a sum of 7, for each value of the face on one die, there is exactly one valid value for the other die, so the probability is $\boxed{\frac{1}{6}}$.

20. We want the center of the circle to lie within distance $\frac{1}{2}$ of a lattice point (see Fig. 8.1). Tile the plane with unit squares whose centers are lattice points and whose sides are parallel to the axes. The valid points make up a circle of radius $\frac{1}{2}$, and thus area $\frac{\pi}{4}$, inside each unit square, so the probability is $\boxed{\frac{\pi}{4}}$. (This glosses over the fact that we cannot actually choose a point on the plane with uniform probability, but this is the answer you get if, for example, you choose the center within a square on the plane and then take the limit as the square's side goes to infinity.)

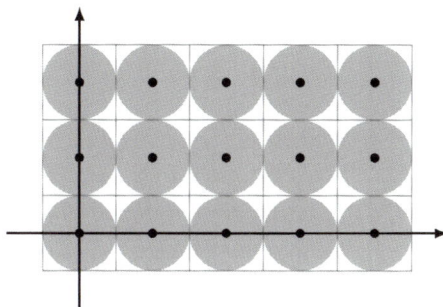

Figure 8.1

21. $A \to B$ is false if and only if A is true and B is false, so the probability that it is true is $1 - \frac{\sqrt{3}}{4}\left(1 - \frac{1}{\sqrt{3}}\right) = 1 - \frac{\sqrt{3}-1}{4} = \boxed{\frac{5-\sqrt{3}}{4}}$.

22. **Solution 1:** We know that 70 is a weighted average of 74 (with weight p) and 58 (with weight $1 - p$); thus, $74p + 58(1 - p) = 70$, so $p = \frac{70-58}{74-58} = \boxed{\frac{3}{4}}$.

    Solution 2: Let x be the number of students who passed and y be the number who failed. The sum of all the grades in the class is $74x + 58y$ and there are $x + y$ students, so the overall average grade is $\frac{74x+58y}{x+y} = 70$. Thus $74x + 58y = 70x + 70y$, or $x = 3y$, and the fraction of students who passed is $\frac{x}{x+y} = \boxed{\frac{3}{4}}$.

23. The probability of picking an even number is $\frac{1}{2^2} + \frac{1}{2^4} + \cdots = \frac{\frac{1}{2^2}}{1 - \frac{1}{2^2}} = \frac{1}{3}$. The probability of picking an even multiple of 5 (i.e., a multiple of 10) is $\frac{1}{2^{10}} + \frac{1}{2^{20}} + \cdots = \frac{\frac{1}{2^{10}}}{1 - \frac{1}{2^{10}}} = \frac{1}{1023}$. So the probability picking of an even number that is not a multiple of 5 is $\frac{1}{3} - \frac{1}{1023} = \boxed{\frac{340}{1023}}$.

24. The units digits of successive powers of 3 cycle through 3, 9, 7, and 1, and those of 7 through 7, 9, 3, and 1; thus, in 100 consecutive powers, there are an equal number of each. There are

three combinations of these units digits that have 4 as the units digit of their sum: $1 + 3$, $3 + 1$, and $7 + 7$. Thus the probability is $\frac{3}{4 \cdot 4} = \boxed{\frac{3}{16}}$.

25. **Solution 1:** The probability of rolling a 9 on any one throw is $\frac{4}{36} = \frac{1}{9}$. For Lucy to win, the first 9 must occur on the $(3k + 2)^{\text{th}}$ roll for k equal to one of $0, 1, \ldots$; for this to happen, the rolls need to consist of $3k + 1$ rolls of anything besides 9, and then a 9. The probability of this happening for any particular k is $\left(\frac{8}{9}\right)^{3k+1} \frac{1}{9}$, and so the total probability of this happening for any value of k is

$$\sum_{k=0}^{\infty} \left(\frac{8}{9}\right)^{3k+1} \frac{1}{9} = \frac{8}{81} \sum_{k=0}^{\infty} \left(\frac{512}{729}\right)^k = \frac{\frac{8}{81}}{1 - \frac{512}{729}} = \boxed{\frac{72}{217}}.$$

Solution 2: The probability of Lucy winning on her first throw is $p = \frac{8}{9} \cdot \frac{1}{9} = \frac{72}{729}$, and the probability of anyone else winning on their first throw is $q = \frac{1}{9} + \left(\frac{8}{9}\right)^2 \frac{1}{9} = \frac{145}{729}$. If nobody wins on their first throw, then we are effectively back where we started: Lucy winning on her next roll is a success and anyone else winning on their next roll is a failure, and so on. So we can assume that someone wins on their first roll, and so the probability that it is Lucy is $\frac{p}{p+q} = \frac{72}{72+145} = \boxed{\frac{72}{217}}$.

In general, suppose we have a trial that succeeds with probability p, fails with probability q, and requires a retry with probability $1 - p - q$. Let the overall probability of a success be P; a success can come from either a success on the first trial, or a retry and then a success on some later trial. Thus $P = p + (1 - p - q)P$, so $(p + q)P = p$ and $P = \frac{p}{p+q}$. Or, using conditional probabilities: getting a retry is like having done nothing at all, so we can ignore retries and the overall probability of a success is $P(\text{trial is a success} \mid \text{trial is not a retry}) = \frac{p}{p+q}$.

26. **Solution 1:** Imagine laying out all the letters from left to right in the order that they are drawn and numbering them from 1 to 7. We only care about the positions of the two A's; there are $\binom{7}{2} = 21$ different possibilities. We want the leftmost A to be in position 1, 3, or 5. If the leftmost A is in position 5, the other A must be in

one of the 2 positions after that; similarly, if the leftmost A is in position 3, there are 4 possibilities for the other one, and position 1 leads to 6 possibilities. Thus the probability is $\frac{2+4+6}{21} = \boxed{\frac{4}{7}}$.

Solution 2: Elle can win on the 1st, 3rd, or 5th pick. To win on the 1st pick, she must pick an A immediately, with probability $\frac{2}{7}$. To win on the 3rd pick, she must pick something besides an A, with probability $\frac{5}{7}$. Then we want Warner to pick something besides an A; there are two A's and four other letters left, so the probability is $\frac{4}{6}$, and Elle to pick an A after that, with probability $\frac{2}{5}$. Similar logic applies to the 5th pick, so the total probability is

$$\frac{2}{7} + \frac{5}{7} \cdot \frac{4}{6} \cdot \frac{2}{5} + \frac{5}{7} \cdot \frac{4}{6} \cdot \frac{3}{5} \cdot \frac{2}{4} \cdot \frac{2}{3} = \frac{12}{21} = \boxed{\frac{4}{7}}.$$

27. The only outcomes that they can both have are 2, 3, 4, 5, and 6. For Helen, the probability of each outcome is $\frac{1}{6}$; for Jim, the probability of outcome n is $\frac{n-1}{36}$ (for these values), so the total probability is

$$\frac{1}{6}\left(\frac{1}{36} + \frac{2}{36} + \frac{3}{36} + \frac{4}{36} + \frac{5}{36}\right) = \frac{15}{216} = \boxed{\frac{5}{72}}.$$

28. **Solution 1:** Each resident can have one of eight birthdays, leading to $8^3 = 512$ combinations for the three birthdays collectively. There are $\binom{8}{3} = 56$ ways to choose three distinct birthdays, and $56 \cdot 3! = 336$ ways to assign them to the people. Thus the probability of having any shared birthday is $1 - \frac{336}{512} = \boxed{\frac{11}{32}}$.

Solution 2: Consider the three people in order. The first can have any birthday. The probability that the second's birthday is different is $\frac{7}{8}$, and, if it is, the probability that the third has a different birthday from both of them is $\frac{6}{8}$. Thus the probability that all three have different birthdays is $\frac{7}{8} \cdot \frac{6}{8} = \frac{21}{32}$, so the probability that at least two have the same birthday is $1 - \frac{21}{32} = \boxed{\frac{11}{32}}$.

29. Any number satisfying the condition must be a multiple of $\text{lcm}(2,3,11) = 66$, so it must be of the form $66k$, with $k \geq \lceil \frac{100}{66} \rceil = 2$ and $k \leq \lfloor \frac{1000}{66} \rfloor = 15$, which gives $15 - 2 + 1 = 14$ values of k to consider. However, if k is a multiple of 2^2 or 3^2, then $66k$ is a multiple of 2^3 or 3^3, respectively. Thus k cannot equal 4, 8, 9, or 12, which leaves 10 values. Meanwhile, there are $1000 - 100 - 1 = 899$ integers

in total between 100 and 1000 exclusive, so the probability is $\boxed{\frac{10}{899}}$.
(We would also need to exclude multiples of 11^2 and perfect cubes, but there are not any of those in the range to begin with.)

30. **Solution 1:** Let p_n be the probability of first getting a 6 on the $2n$th roll; for this to happen, we need to get anything besides 6 for $2n-1$ rolls and then roll a 6, so $p_n = \left(\frac{5}{6}\right)^{2n-1}\frac{1}{6} = \frac{5^{2n-1}}{6^{2n}} = \frac{1}{5}\left(\frac{25}{36}\right)^n$.
All of these events are mutually exclusive, so the probability of first getting a 6 on any even-numbered roll is

$$\sum_{i=1}^{\infty} p_i = \frac{1}{5}\sum_{i=1}^{\infty}\left(\frac{25}{36}\right)^n = \frac{1}{5}\frac{\frac{25}{36}}{1-\frac{25}{36}} = \boxed{\frac{5}{11}}.$$

Solution 2: With probability $q = \frac{1}{6}$, the first roll is a 6, so we have failed; with probability $p = \frac{5}{6}\cdot\frac{1}{6} = \frac{5}{36}$, the second roll is a 6 and the first is not, so we have succeeded. As described in the solution to problem 25, the total probability of a success is therefore
$\frac{p}{p+q} = \frac{\frac{5}{36}}{\frac{5}{36}+\frac{1}{6}} = \boxed{\frac{5}{11}}$.

Level 3

31. There are $\binom{6}{3} = 20$ ways of getting exactly 3 heads in 6 flips, only one of which has no heads in the first 3 flips (the one where the last 3 are heads). There are $2^6 = 64$ possibilities for the outcome of all 6 flips, of which $2^3 = 8$ have no heads in the first 3 flips (since the first three must be tails, but the last three can be either). Thus the probability is $\frac{20-1}{64-8} = \boxed{\frac{19}{56}}$.

32. The probability of getting any particular sequence of n heads and tails on the first n flips of the coin is $\frac{1}{2^n}$, no matter what the sequence is. The statement that $b = 4a$ is simply a requirement that the first $4a$ flips of the coin are a certain sequence (namely, the ath flip and the $4a$th flip are tails, and no others are), so the probability of it being true is $\left(\frac{1}{2}\right)^{4a} = \left(\frac{1}{16}\right)^a$. No sequence of flips can satisfy it for two different values of a, so the total probability

of it being true for any value of a is

$$\sum_{a=1}^{\infty}\left(\frac{1}{16}\right)^{a} = \frac{\frac{1}{16}}{1 - \frac{1}{16}} = \boxed{\frac{1}{15}}.$$

33. As problem 25 shows, we can simply treat Felicia as having a four-sided die. For Patrick to win on his first roll, we need to have Felicia and Nathan both roll something besides 1 first, so the probability it happens is $\frac{3}{4} \cdot \frac{5}{6} \cdot \frac{1}{12} = \frac{15}{288}$. If Patrick wins on the second trial, then no one won on the first trial (which happens with probability $\left(\frac{3}{4} \cdot \frac{5}{6} \cdot \frac{11}{12}\right) = \frac{165}{288}$), and then Patrick is the first to roll a 1 (which happens with probability $\frac{15}{128}$). Thus, the probability that Patrick wins on the second trial is $\frac{165}{288} \cdot \frac{15}{288}$. Continuing, for Patrick to win on the third trial, no one won on the first two trials and then Patrick won, which happens with probability $\left(\frac{165}{288}\right)^{2} \cdot \frac{15}{288}$. Thus the probability that Patrick wins is $\frac{15}{288} + \frac{165}{288} \cdot \frac{15}{288} + \left(\frac{165}{288}\right)^{2} \cdot \frac{15}{288} + \cdots$ which is an infinite geometric series whose sum is:

$$\frac{a}{1-r} = \frac{\frac{15}{288}}{1 - \frac{165}{288}} = \frac{15}{123} = \boxed{\frac{5}{41}}.$$

34. Suppose Chris sits down first. There are 16 places for John to sit after him, and 15 for Kim after that; if we exclude the seats next to Chris, then there are instead 14 and 13. The probability that neither of them is next to Chris is therefore $\frac{14 \cdot 13}{16 \cdot 15} = \frac{91}{120}$, and so the probability that at least one of them is next to Chris is $1 - \frac{91}{120} = \boxed{\frac{29}{120}}$.

35. The probability of getting tails on the ith throw is $1 - \frac{1}{2^i} = (2^i - 1) \cdot \frac{1}{2^i}$, so the probability of getting that and also heads on all the other throws is $(2^i - 1) \cdot \frac{1}{2^1} \cdot \frac{1}{2^2} \cdots \frac{1}{2^{10}} = \frac{2^i - 1}{2^{55}}$. Thus the probability, multiplied by 2^{55}, of getting exactly one tail on any of the throws is

$$\sum_{i=1}^{10} 2^i - 1 = -10 + \sum_{i=1}^{10} 2^i = -10 + \left(2^{11} - 2\right) = \boxed{2036}.$$

Chapter 9

Functional Equations

Questions

Level 1

1. If f is a function defined on the real numbers such that $f(x+y) + f(x-y) = 2f(x)f(y)$ and $f(1) > 0$, compute $f(0)$.
2. If $f(a) + f(b) = f(ab)$ for all real a and b, compute $f(1999)$.
3. If $f(x) = \frac{1}{\sqrt{1-x}}$ and $f(a) = 2$, compute $f(1-a)$.
4. If $f(x) = \frac{3}{x+1}$ and $f(g(x)) = \frac{6-3x}{6-x}$, then $g(x)$ may be expressed in the form $g(x) = \frac{a}{x+b}$. Compute ab.
5. For all real x, $f(x) = 5x$ and $g(f(x)) = x$. Compute $g(2004)$.
6. If $f(x) = 8x - 3$, find the ordered pair of real numbers (c, d) such that $f(d) = 10c$ and $f(c) = d$.
7. If, for all x, $f(x) = 2x - 7$ and $f(g(x)) = x$, express $g(x)$ in terms of x.
8. For all real x, $f(x) = x + 3$, and $g(x)$ is a polynomial of degree 2 such that $g(f(x)) = x^2 + 2$. Find $g(x)$ in terms of x.
9. If $f(x) = \frac{x}{2} - 3$, compute the numerical value of $f(f(f(8)))$.
10. If $f(x) = \sqrt{x-7}$, find all real numbers x such that $f(f(x)) = 3$.
11. If $f(x-1) = x^2 + 3x + 2$, find $f(x+1)$.
12. If $f(x) = \frac{1}{1-x}$ and $g(x) = 1 - \frac{1}{x}$ for all $x > 1$, find the numerical value of $f(g(1990)) - g(f(1990))$.
13. For all real x, $f(2x) = x^2 - x + 3$. Express $f(x)$ in terms of x.
14. For all complex numbers z such that $z^2 \neq 1$, $f(z) = \frac{z^2+1}{z^2-1}$. If i represents the imaginary unit, compute $f(i)$.

157

Level 2

15. If $f(2) = 5$ and $f(x)f(x+1) = 3$ for all real x, find $f(10)$.

16. If $f(x) = -\frac{1}{1+x}$ and $g(x) = -1 - x$, find the numerical value of $f(g(f(g(1987))))$.

17. If $f(x)$ is a polynomial in x such that $f(1-x) + 2f(x) = x$, find $f(x)$.

18. Let $f(x) = 1 - x$ and $g(x) = \frac{1}{x}$. Find all real values of y such that $f(g(f(g(y)))) = 3$.

19. Suppose $f(1) = 1$ and, for all positive integers m and n,

$$f(m+n) = \frac{f(m)f(n)}{f(m) + f(n)}.$$

Find $f(1987)$.

20. Suppose that $f(1) = 1$ and for all positive integers n,

$$f(2n) = 2f(n) - 1 \quad \text{and} \quad f(2n+1) = 2f(n) + 1.$$

Compute $f(100)$.

21. If $f(x) = 1990 - 5x$, how many ordered pairs of integers (a, b) are there such that $a \in [0, 1000]$, $b \in [0, 1000]$, and $f(2a) = f(a+b) - f(a-b)$?

22. Suppose $f(m, n)$ satisfies the following conditions for all positive integers m and n:

- $f(m, n) = f(m+n, n)$,
- $f(m, n) = f(n, m)$, and
- $f(m, m) = m$.

Compute $f(89, 55)$.

23. Suppose $f(x, y)$ is a function such that $f(x, 1) = x^2 + 59x$ and $f(x - 3, y - 4) = f(x, y)$. Find $f(93, 93)$.

24. If $2f(x) + f\left(\frac{1}{x}\right) = \frac{x}{2}$ for any nonzero real number x, compute $f(3)$.

25. Compute $f(2)$ if $2f(x) + \sqrt{2}f\left(\frac{1}{x}\right) = 2^x$ for all nonzero real x.

26. Compute the number of radians in $f(1) + f(2) + f(3)$ if $f(x) = \arctan \frac{1}{x}$.

27. Compute $f(4)$ if $2f(x^2) + 3f(19 - 5x) = x^3$ for all x.

28. If $\frac{3f(x)}{f(x)-3} = x$ for all real numbers x, compute $f(6)$.

29. If $f_1(x) = \frac{x\sqrt{3}-1}{x+\sqrt{3}}$, $f_2(x) = f_1(f_1(x))$, $f_3(x) = f_1(f_2(x))$, and, in general, $f_n(x) = f_1(f_{n-1}(x))$ for $n > 1$, compute $f_{1986}(1985)$.

Level 3

30. Let $a = \frac{1}{2}(-1 + \sqrt{4\sqrt[3]{3} - 7})$ and let $f(x) = (a^2 + a + 2)^x$. Compute $f(f(f(f(f(3)))))$.

31. For any positive integers a and b, let $f(a, b)$ represent the number of positive perfect a^{th} powers divisible by a and less than or equal to b. Compute

$$\sum_{n=2}^{\infty} f(n, 10^5).$$

32. If $f(x) = x^{1990}$, compute the remainder when $f(1) + f(2) + \cdots + f(1990)$ is divided by 5.

33. Suppose $f(x)$ is defined for all positive x such that $11f(x + 1) + 5f(x^{-1} + 1) = \log_{10} x$. Compute $f(6) + f(17) + f(126)$.

Answers

Level 1

1. Letting $x = 1$ and $y = 0$ gives $2f(1) = 2f(1)f(0)$; since we know $f(1) \neq 0$, we must have $f(0) = \boxed{1}$.

2. Letting $a = 1999$ and $b = 0$ gives $f(1999) + f(0) = f(0)$, so $f(1999) = \boxed{0}$.

3. We have $\frac{1}{\sqrt{1-a}} = 2$, so $1 - a = \frac{1}{4}$ and $f(1-a) = \frac{1}{\sqrt{\frac{3}{4}}} = \sqrt{\frac{4}{3}} = \boxed{\frac{2\sqrt{3}}{3}}$.

4. $\frac{3x-6}{x-6} = f(g(x)) = \frac{3}{\frac{a}{x+b}+1} = \frac{3x+3b}{x+a+b}$, so $3b = -6$ and $a + b = -6$, so $b = -2$, $a = -4$, and $ab = \boxed{8}$.

5. $g(5x) = x$, so we can replace x by $\frac{x}{5}$ and $g(x) = \frac{x}{5}$, so $g(2004) = \boxed{\frac{2004}{5}}$.

6. The given equations become $8d - 3 = 10c$ and $8c - 3 = d$; isolating the 3 in each gives $3 = 8d - 10c = 8c - d$, so $9d = 18c$ or $d = 2c$. Substituting into the first equation gives $16c - 3 = 10c$, so $c = \frac{1}{2}$, $d = 1$, and $(c, d) = \boxed{(\frac{1}{2}, 1)}$.

7. $2g(x) - 7 = x$, so $2g(x) = x + 7$ and $g(x) = \boxed{\frac{x+7}{2}}$.

8. $g(x+3) = x^2 + 2$; substitute $x - 3$ for x to get $g(x) = (x-3)^2 + 2 = \boxed{x^2 - 6x + 11}$.

9. Compute: $f(8) = 1$, $f(f(8)) = f(1) = -\frac{5}{2}$, and $f(f(f(8))) = f\left(-\frac{5}{2}\right) = \boxed{-\frac{17}{4}}$.

10.

$$\sqrt{\sqrt{\sqrt{x-7}-7}} = 3$$

$$\sqrt{\sqrt{x-7}-7} = 9$$

$$\sqrt{x-7} = 16$$

$$x - 7 = 256$$

$$x = \boxed{263}.$$

11. Note $f(x-1) = (x+1)(x+2)$ and substitute $x + 2$ for x to get $f(x+1) = (x+3)(x+4) = \boxed{x^2 + 7x + 12}$.

12. $f(g(x)) = \frac{1}{1 - \left(1 - \frac{1}{x}\right)} = x$ and $g(f(x)) = 1 - \frac{1}{\frac{1}{1-x}} = x$, so for any x, $f(g(x)) - g(f(x)) = \boxed{0}$.

13. Substitute $\frac{x}{2}$ for x to get $f(x) = \left(\frac{x}{2}\right)^2 - \frac{x}{2} + 3 = \boxed{\frac{x^2}{4} - \frac{x}{2} + 3}$.

14. By definition, $i^2 = -1$, and so $f(i) = \frac{-1+1}{-1-1} = \boxed{0}$.

Level 2

15. We have $f(x+1) = \frac{3}{f(x)}$ and, substituting $x + 1$ for x, $f(x+2) = \frac{3}{f(x+1)} = f(x)$. Thus $f(2) = f(4) = \cdots = f(10) = \boxed{5}$.

16. $f(g(x)) = -\frac{1}{1+(-1-x)} = -\frac{1}{-x} = \frac{1}{x}$, so $f(g(f(g(1987)))) = f\left(g\left(\frac{1}{1987}\right)\right) = \boxed{1987}$.

17. Suppose the degree of $f(x)$ is n and its highest-order term is cx^n. Then the term of highest possible order of $f(1-x) + 2f(x)$ is $(2 + (-1)^n)cx^n$. Since $2 + (-1)^n \neq 0$, this must equal x, so $n = 1$ and $c = 1$. So $f(x) = x + b$; we have $x = ((1-x) + b) + 2x + 2b = x + (3b+1)$. Thus $b = -\frac{1}{3}$ and $f(x) = \boxed{x - \frac{1}{3}}$.

18. $f(g(y)) = 1 - \frac{1}{y}$ and $3 = f(g(f(g(y)))) = 1 - \frac{1}{1-\frac{1}{y}} = 1 - \frac{y}{y-1} = -\frac{1}{y-1}$, so $y - 1 = -\frac{1}{3}$ and $y = \boxed{\frac{2}{3}}$.

19. Let $n = 1$ so that $f(m+1) = \frac{f(m)}{1+f(m)}$. $f(2) = \frac{1}{1+1} = \frac{1}{2}, f(3) = \frac{\frac{1}{2}}{1+\frac{1}{2}} = \frac{1}{3}$. Thus by observation or induction, $f(k) = \frac{1}{k}$. So $f(1987) = \boxed{\frac{1}{1987}}$.

20. **Solution 1:** The first equation implies that $f(n) = 2f\left(\frac{n}{2}\right) - 1$ if n is even, and the second that $f(n) = 2f\left(\frac{n-1}{2}\right) + 1$ if n is odd. Thus,
$$f(100) = 2f(50) - 1 = 2(2f(25) - 1) - 1$$
$$= 4f(25) - 3 = 4(2f(12) + 1) - 3$$
$$= 8f(12) + 1 = 8(2f(6) - 1) + 1$$
$$= 16f(6) - 7 = 16(2f(3) - 1) - 7$$
$$= 32f(3) - 23 = 32(2f(1) + 1) - 23$$
$$= 96 - 23 = \boxed{73}.$$

(The following solution, lemmas and proofs are intended for the advanced reader.)

Solution 2: By induction on the number of base-2 digits in n, we can see that $f(n) = n - \neg n$, where $\neg n$ is the integer produced by replacing all 0s with 1s and vice versa in the base-2 representation of n (see Lemma 1 below). But, if we let $T = 2^{1+\lfloor \log_2 n \rfloor}$ (i.e., the smallest power of 2 greater than n), then $n + \neg n = T - 1$ (see Lemma 2 below). Thus $f(n) = n - \neg n = 2n - T + 1$. If $n = 100$, then $T = 128$, and so $f(100) = 200 - 128 + 1 = \boxed{73}$.

Lemma 1. $f(n) = n - \neg n$. For $n = 1$, we have $\neg 1 = 0$ and $f(1) = 1 - 0 = 1$.

Proof 1. For any $n > 1$, let d be the lowest base-2 digit in n and let m be the number obtained by truncating that digit; then $n = 2m + d$. Similarly, $\neg n = 2 \cdot \neg m + (1 - d)$. We can combine the two given equations into $f(n) = 2f(m) + (2d - 1)$, and so
$$f(n) = 2f(m) + (2d - 1)$$
$$= 2(m - \neg m) + d + (d - 1)$$
$$= (2m + d) - (2 \cdot \neg m + (1 - d))$$
$$= n - \neg n.$$

■

Proof 2. Alternatively, considering building up the argument to f by looking at its base-2 digits from left to right. Let D denote a "digit" with a value of -1. For any n, appending a 0 to n (in base 2) means changing n into $2n$, and therefore changing $f(n)$ into $2f(n) - 1$. But that is the same as appending D to $f(n)$. Similarly, appending 1 to n means appending 1 to $f(n)$. We start with $n = f(n) = 1$, so we have, for example, the following values:

n	$f(n)$
1	1
10	$1D$
100	$1DD$
1000	$1DDD$
10001	$1DDD1$

and, in general, $f(n)$ can be obtained from n by replacing all 0s with Ds. We can evaluate that number by, instead of doing the replacement, subtracting from n a new number which has 1s exactly where n has 0s. But that number is $\neg n$. ∎

Lemma 2. $n + \neg n = 2^{1 + \lfloor \log_2 n \rfloor} - 1.$

Proof. In each digit position, exactly one of n and $\neg n$ has a 1 and the other has a 0; thus adding them involves no carries and their sum is simply a string of as many 1s as n has digits. A string of d 1s has value $\sum_{i=0}^{d-1} 2^i = 2^d - 1$. In general, n has $1 + \lfloor \log_b n \rfloor$ digits when written in base b, so here we have $1 + \lfloor \log_2 n \rfloor$ digits, and the value of the sum is $2^{1 + \lfloor \log_2 n \rfloor} - 1$. ∎

21. Use the definition of $f(x)$ to get $1990 - 10a = (1990 - 5a - 5b) - (1990 - 5a + 5b) = -10b$, or $b = a - 199$. So there is exactly one possible value of a if and only if $b \in [0, 1000 - 199]$, which leads to $1000 - 199 + 1 = \boxed{802}$ possibilities.

22. **Solution 1:** Apply the first two conditions repeatedly until we can apply the third:

$$f(89, 55) = f(34 + 55, 55) = f(34, 55) = f(55, 34)$$

$$= f(34, 21) = f(21, 13) = f(13, 8) = f(8, 5)$$

$$= f(5, 3) = f(3, 2) = f(2, 1) = f(1, 1) = \boxed{1}.$$

Solution 2: Notice that $f(m, n) = \gcd(m, n)$ (evaluating f is simply executing the Euclidean algorithm), and $\gcd(89, 55) = \boxed{1}$.

23. Applying the second equality repeatedly gives $f(x, y) = f(x - 3n, y - 4n)$ for any integer n; $93 = 4 \cdot 23 + 1$, so $f(93, 93) = f(93 - 3 \cdot 23, 1) = f(24, 1) = 24^2 + 24 \cdot 59 = \boxed{1992}$.

24. Let $a = f(3)$ and $b = f\left(\frac{1}{3}\right)$; substituting $x = 3$ and $x = \frac{1}{3}$ into the given equation gives $2a + b = \frac{3}{2}$ and $2b + a = \frac{1}{6}$. Subtracting these gives $a - b = \frac{3}{2} - \frac{1}{6} = \frac{4}{3}$ and adding them gives $3a + 3b = \frac{3}{2} + \frac{1}{6} = \frac{10}{6}$ or $a + b = \frac{5}{9}$, so $a = \frac{\frac{4}{3} + \frac{5}{9}}{2} = \boxed{\frac{17}{18}}$.

25. Let $a = f(2)$ and $b = f\left(\frac{1}{2}\right)$; substituting $x = 2$ and $x = \frac{1}{2}$ into the given equation gives $2a + \sqrt{2}b = 4$ and $2b + \sqrt{2}a = \sqrt{2}$. Subtracting these gives $(2 - \sqrt{2})(a - b) = 4 - \sqrt{2}$, so $a - b = \frac{4 - \sqrt{2}}{2 - \sqrt{2}} = 3 + \sqrt{2}$, and adding them gives $(2 + \sqrt{2})(a + b) = 4 + \sqrt{2}$, so $a + b = \frac{4 + \sqrt{2}}{2 + \sqrt{2}} = 3 - \sqrt{2}$. Finally, $a = \frac{(3 + \sqrt{2}) + (3 - \sqrt{2})}{2} = \boxed{3}$.

26. We want $\arctan 1 + \arctan \frac{1}{2} + \arctan \frac{1}{3}$. We know that $\arctan 1 = \frac{\pi}{4}$, and

$$\tan\left(\arctan \frac{1}{2} + \arctan \frac{1}{3}\right) = \frac{\frac{1}{2} + \frac{1}{3}}{1 - \frac{1}{2} \cdot \frac{1}{3}} = 1,$$

so $\arctan \frac{1}{2} + \arctan \frac{1}{3} = \frac{\pi}{4}$, and the sum of all three is $\boxed{\frac{\pi}{2}}$.

27. Let $a = f(4)$ and $b = f(9)$; substituting $x = 2$ and $x = 3$ into the given equation gives $2a + 3b = 8$ and $2b + 3a = 27$. Subtracting these gives $a - b = 19$ and adding them gives $5a + 5b = 35$ or $a + b = 7$, so $a = \frac{19 + 7}{2} = \boxed{13}$.

28. **Solution 1:** Substitute $x = 6$ and invert both sides of the equation to get $\frac{1}{6} = \frac{f(6) - 3}{3f(6)} = \frac{1}{3} - \frac{1}{f(6)}$ and so $\frac{1}{f(6)} = \frac{1}{3} - \frac{1}{6} = \frac{1}{6}$, so $f(6) = \boxed{6}$.
Solution 2:

$$\frac{3f(6)}{f(6) - 3} = 6$$

$$3f(6) = 6f(6) - 18$$

$$3f(6) = 18$$

$$f(6) = \boxed{6}.$$

29. **Solution 1:**
$$f_2(x) = \frac{(x\sqrt{3}-1)\sqrt{3}-(x+\sqrt{3})}{(x\sqrt{3}-1)+(x+\sqrt{3})\sqrt{3}} = \frac{2x-2\sqrt{3}}{2x\sqrt{3}+2} = \frac{x-\sqrt{3}}{x\sqrt{3}+1},$$

$$f_3(x) = \frac{(x-\sqrt{3})\sqrt{3}-(x\sqrt{3}+1)}{(x-\sqrt{3})+(x\sqrt{3}+1)\sqrt{3}} = \frac{-4}{4x} = -\frac{1}{x},$$

$$f_6(x) = -\frac{1}{-\frac{1}{x}} = x.$$

Thus $f_{6k}(x) = f_6(f_{6(k-1)}(x)) = f_{6(k-1)}(x) = \cdots = f_6(x) = x$ for any positive integer k. 1986 is a multiple of 6 and so $f_{1986}(1985) = \boxed{1985}$.

Solution 2: For any 2×2 matrix $M = \begin{pmatrix} a & b \\ c & d \end{pmatrix}$, let $g[M](x) = \frac{ax+b}{cx+d}$. Note that $g[M](x) = g[tM](x)$ for any scalar t. Some computation shows that, if N is another 2×2 matrix, then $g_M(g_N(x)) = g[MN](x)$. If we let $M = \begin{pmatrix} \sqrt{3} & -1 \\ 1 & \sqrt{3} \end{pmatrix}$, then $f_1(x) = g_M(x)$ and, by induction, $f_k(x) = g[M^k](x)$ for all k. You can tell by looking that M is proportional to the rotation matrix corresponding to some multiple of $30°$; since 1986 is a multiple of 6, M^{1986} is proportional to the rotation matrix corresponding to some multiple of $180°$, so it is proportional to the identity matrix I (possibly with a negative ratio, but that does not matter). But $g[I](x) = \frac{1x+0}{0x+1} = x$, so $f_{1986}(1985) = \boxed{1985}$.

Level 3

30. Let $b = \frac{1}{2}(-1 - \sqrt{4\sqrt[3]{3}-7})$. Then $a + b = -1$ and $ab = 2 - \sqrt[3]{3}$, so a and b are the roots of the equation $x^2 + x + 2 - \sqrt[3]{3} = 0$. Thus $a^2 + a + 2 = \sqrt[3]{3}$, so $f(3) = (\sqrt[3]{3})^3 = 3$, and $f(f(\cdots f(3) \cdots)) = \boxed{3}$.

31. Consider each n.

   - $n = 2$: a 2nd power divisible by 2 is the square of an even number. We have $\sqrt{10^5} = 100\sqrt{10} = 316.2 \cdots$, so there are $\lfloor \frac{316}{2} \rfloor = 158$ such numbers less than or equal to 10^5.
   - $n = 3$: a 3rd power divisible by 3 is the cube of a multiple of 3. We have $\sqrt[3]{10^5} = 10\sqrt[3]{100}$ and $4^3 < 100 < 5^3$, so $40 < 10\sqrt[3]{100} < 50$. It turns out that $45^3 < 10^5 < 48^3$, so there are $\frac{45}{3} = 15$ such numbers less than or equal to 10^5.
   - $n = 4$: a 4th power divisible by 4 is the 4th power of an even number. We have $\sqrt[4]{10^5} = \sqrt{\sqrt{10^5}} = \sqrt{316.2 \cdots}$ and $16^2 <$

$316.2 \cdots < 18^2$, so there are $\frac{16}{2} = 8$ such numbers in the range.

- $n = 5$: a 5th power divisible by 5 is the 5th power of a multiple of 5. There are 2 such numbers (5^5 and 10^5) less than or equal to 10^5.
- $n = 6$: a 6th power divisible by 6 is the 6th power of a multiple of 6. We have $6^6 = 46656 < 10^5$ and $12^6 > 10^5$, so there is 1 such number less than or equal to 10^5.
- $n = 7$: a 7th power divisible by 7 is the 7th power of a multiple of 7. We have $7^7 = 7^4 \cdot 7^3 = 2401 \cdot 343 > 10^5$, so there are no such numbers less than or equal to 10^5, and the same holds for $n > 7$.

Thus the sum is $158 + 15 + 8 + 2 + 1 = \boxed{184}$.

32. Modulo 5,

$$0^{1990} \equiv 0;$$

$$1^{1990} \equiv (1^2)^{995} \equiv 1^{995} \equiv 1;$$

$$2^{1990} \equiv (2^2)^{995} \equiv (-1)^{995} \equiv -1;$$

$$3^{1990} \equiv (3^2)^{995} \equiv (-1)^{995} \equiv -1;$$

$$4^{1990} \equiv (4^2)^{995} \equiv 1^{995} \equiv 1 \qquad .$$

Since $0 + 1 - 1 - 1 + 1 = 0$, the sum of the values of f applied to any 5 consecutive integers is a multiple of 5. Since 1990 is itself a multiple of 5, we can break the whole sum up into such sums, so the whole thing is a multiple of 5 and the remainder is $\boxed{0}$.

33. Let $a = f(x + 1)$ and $b = f(x^{-1} + 1)$; the given equation is $11a + 5b = \log_{10} x$, and substituting $\frac{1}{x}$ for x gives $11b + 5a = -\log_{10} x$. Subtracting these gives $6(a - b) = 2\log_{10} x$ and adding them gives $16(a + b) = 0$, so $a = -b$, $12a = 2\log_{10} x$, and $a = \frac{1}{6}\log_{10} x$. Thus, $f(x) = \frac{1}{6}\log_{10}(x - 1)$ and

$$f(6) + f(17) + f(126) = \frac{1}{6}(\log_{10} 5 + \log_{10} 2^4 + \log_{10} 5^3)$$

$$= \frac{1}{6}\log_{10} 10^4$$

$$= \boxed{\frac{2}{3}}.$$

Appendix

Algebra Facts

A.1.
$$(a+b)^3 = a^3 + 3a^2b + 3ab^2 + b^3,$$
$$(a-b)^3 = a^3 - 3a^2b + 3ab^2 - b^3,$$

from above:

$$a^3 + b^3 = (a+b)^3 - 3ab(a+b),$$
$$a^3 - b^3 = (a-b)^3 - 3ab(a+b).$$

A.2. **The Binomial Theorem:**

$$(a+b)^n = {}_nC_0 a^n + {}_nC_1 a^{n-1}b + {}_nC_2 a^{n-2}b^2$$

$$+ \cdots +_n C_{n-1}a^1b^{n-1} + {}_nC_n b^n = \sum_{k=0}^{n} {}_nC_k a^{n-k}b^k,$$

where ${}_nC_k = \frac{n!}{k!(n-k)!}$. In particular, ${}_nC_0 = {}_nC_n = 1$.

A.3.
$$a^n - b^n = (a-b)(a^{n-1} + a^{n-2}b^2 + a^{n-3}b^3 + \cdots + b^{n-1}),$$
$$a^n + b^n = (a+b)(a^{n-1} - a^{n-2}b^2 + a^{n-3}b^3 - \cdots + b^{n-1})$$

when n is odd.

When n is even, this is not factorable with real coefficients.

A.4. **Vieta's Formulas:** For the nth degree polynomial

$$p(x) = a_n x^n + a_{n-1}x^{n-1} + a_{n-2}x^{n-2} + \cdots + a_0,$$

the sum of the product of the roots taken two at a time is $\frac{a_{n-2}}{a_n}$. The sum of the product of the roots taken three at a time is $-\frac{a_{n-3}}{a_n}$, etc. The product of the roots is $(-1)^n \frac{a_0}{a_n}$.

A.5. **The Remainder Theorem:** If a polynomial $p(x)$ is divided by $x-a$, the remainder is $p(a)$.

A.6. **The Factor Theorem:** For the nth degree polynomial $p(x)$, $x = a$ is a root of $p(x) = 0$ if and only if $x - a$ is a factor of $p(x)$.

A.7. If the nth degree polynomial

$$p(x) = a_n x^n + a_{n-1} x^{n-1} + a_{n-2} x^{n-2} + \cdots + a_0,$$

where $p(x)$ has integer coefficients, has a rational root $\frac{p}{q}$ (which we assume is in lowest terms), then p divides a_0 and q divides a_n.

A.8. If $r = a + bi$ is a root of the equation $p(x) = 0$, where $p(x)$ is a polynomial *with real coefficients*, then its conjugate, $s = a - bi$, is also a root of this equation.

A.9. If $p(x)$ is a polynomial with *rational* coefficients and if $r = a + b\sqrt{c}$ is a root of $p(x) = 0$ where a, b, and c are rational, c is positive, but not a perfect square, then its conjugate, $s = a - b\sqrt{c}$, is also a root of this equation.

A.10. If $a^3 + b^3 = (a + b)^3$, then $a = 0$, $b = 0$ or $a = -b$.

A.11. **The Principal of Inclusion/Exclusion:** In the simple case of two finite sets, A and B, we get $|A \cup B| = |A| + |B| - |A \cap B|$ where $|X|$ refers to the number of elements in set X (also called its cardinality). For three sets we would get:

$$|A \cup B \cup C| = |A| + |B| + |C| - |A \cap B| - |A \cap C| - |B \cap C| + |A \cap B \cap C|.$$

This can be extended to any number of sets.

Geometry Facts

G.1. In $\triangle ABC$, $\quad a^2 + b^2 = c^2$ if and only if the triangle is a right triangle with right angle at C.

G.2. The three medians of a triangle meet at a point (called the centroid). That point divides the medians into segments with a ratio of 2:1.

G.3. If m_c is the length of the median to side c of a triangle, then $m_c^2 = \frac{1}{2}(a^2 + b^2) - \frac{1}{4}c^2$.

G.4. If the length of the angle bisector to side c of a triangle is represented by t_c and it divides c into segments with lengths m and n, where m is the length of the segment closer to a and n is the length of the segment closer to b, then

$$\frac{a}{b} = \frac{m}{n} \quad \text{and} \quad ab - mn = t_c^2$$

G.5. If a, b, and c are the measures of the sides of a triangle, K its area, s its semi-perimeter $(s = \frac{a+b+c}{2})$, r the measure of the radius of the inscribed circle and R the measure of the radius of the circumscribed circle, then

$$K = \sqrt{s(s-a)(s-b)(s-c)}$$

$$= rs$$

$$= \frac{abc}{4R}$$

$$= \frac{1}{2}ab\sin C.$$

The first equation above is known as *Heron's* or *Hero's formula*.

G.6. In any equilateral triangle, the sum of the distances from any interior point to the three sides is equal to the length of the altitude of the triangle.

G.7. In any convex parallelogram $ABCD$

$$AC^2 + BD^2 = AB^2 + BC^2 + CD^2 + DA^2.$$

G.8. A quadrilateral is cyclic (that is, can be inscribed in a circle) if and only if opposite angles are supplementary.

G.9. **Ptolemy's Theorem:** If $ABCD$ is a cyclic quadrilateral, then

$$AB \cdot CD + AD \cdot BC = AC \cdot BD.$$

G.10. If the measures of the sides of a cyclic quadrilateral are a, b, c, and d, then the area, K, of the quadrilateral is given by

$$K = \sqrt{s(s-a)(s-b)(s-c)(s-d)},$$

where $s = \frac{a+b+c+d}{2}$. (This is known as *Brahmagupta's formula*.)

G.11. In a circle of radius r, the length of a chord which subtends an inscribed angle of θ is $2r\sin\theta$.

G.12. The area of any equilateral triangle is equal to $\frac{s^2\sqrt{3}}{4}$ where s is the length of a side of the triangle.

G.13. The area of a parallelogram with adjacent sides having lengths a and b is $ab\sin\theta$ where θ is the angle between these sides.

G.14. The area of a rhombus is $\frac{1}{2}$ the product of the lengths of the diagonals.

G.15. The perimeters of similar figures are in the same proportion as the corresponding sides (or any corresponding parts, for example the altitudes drawn to corresponding sides). The areas of similar figures are to each other as the square of the ratio of the corresponding sides (or the square of the ratio of any corresponding parts).

G.16. A median drawn to the side of a triangle divides the triangle in two triangles with equal areas.

G.17. When the three medians of a triangle are drawn, they divide the triangle into six triangles with equal area.

G.18. If the feet of the three medians of a triangle are used to form a new triangle, the area of the new triangle is $\frac{3}{4}$ the area of the original triangle.

G.19. The distance from a point (x_1, y_1) to a line $ax + by + c = 0$ is given by:

$$\frac{|ax_1 + by_1 + c|}{\sqrt{a^2 + b^2}}.$$

The distance from a point (x_1, y_1, z_1) in three-dimensional space to a plane $ax + by + cz + d = 0$ is given by:

$$\frac{|ax_1 + by_1 + cz_1 + d|}{\sqrt{a^2 + b^2 + c^2}}.$$

G.20. If the x, y, and z intercepts of a plane are a, b, and c, respectively, then the equation of the plane is

$$\frac{x}{a} + \frac{y}{b} + \frac{z}{c} = 1.$$

G.21. The median drawn to the hypotenuse of a right triangle has length half the length of the hypotenuse. Conversely, if a median in a triangle is half the length of the side it is drawn to, then the triangle is a right triangle.

G.22. If $(x_1, y_1), (x_2, y_2), (x_3, y_3)$, and (x_4, y_4) are consecutive vertices of a parallelogram, then $x_1 = x_2 - x_3 + x_4$ and $y_1 = y_2 - y_3 + y_4$.

G.23. For any point P inside of a rectangle $ABCD$, $AP^2 + PC^2 = BP^2 + PD^2$.

G.24. The sum of the squares of the lengths of the three sides of a triangle is equal to $\frac{4}{3}$ the sum of the squares of the lengths of the three medians drawn to the three sides.

G.25. **Power of a Point:** Given point P and circle O with radius r, s is the distance from P to the center of the circle, h (called the power of the point) is the relative distance from P to the circle, so:

$$h = s^2 - r^2.$$

(Note this is the square of the length of a tangent from P to the circle.)

The Power of a Point Theorem: In the case above, if a line through P intersects the circle at A and B respectively, then $h = PA \cdot PB$ and h is constant for all such lines.

G.26. **Pythagorean Theorem in Three Dimensions:** The diagonal d of a rectangular solid of sides a, b, c is $d = \sqrt{a^2 + b^2 + c^2}$. If the solid is a cube with side a, we get $d = \sqrt{3a^2} = a\sqrt{3}$.

G.27. **Menelaus' Theorem:** Given a triangle ABC, and a transversal that crosses $\overline{AC}$ at E, $\overline{BC}$ at D, and the extension of $\overline{AB}$ at F, then the following is true:

$$\frac{AF}{BF} \cdot \frac{BD}{CD} \cdot \frac{CE}{AE} = -1,$$

which can also be written $AF \cdot BD \cdot CE = -BF \cdot CD \cdot AE$.

G.28. **Stewarts Theorem:** Given $\triangle ABC$, draw a line segment from A to $\overline{BC}$ (this is known as a Cevian) meeting a at any point E. Let d be the length of the segment. Referring to the below diagram, we get the following formula: $a \cdot (mn + d^2) = b^2 m + c^2 n$.

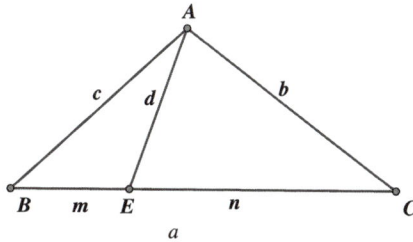

(Note) A Cevian is any line segment in a triangle from a vertex to the opposite side.

G.29. **Chord-Chord Theorem:** If two chords intersect within a circle, then the product of the lengths of the segments of one chord equals the product of the lengths of the segments of the other chord.

Trigonometry Facts

T.1. $\sin \theta = \cos(90 - \theta)$.
T.2. $\sin^2 \theta + \cos^2 \theta = 1$.
T.3. $\tan^2 \theta + 1 = \sec^2 \theta$.
T.4. $\cot^2 \theta + 1 = \csc^2 \theta$.

T.5. $\sin 2\theta = 2\sin\theta\cos\theta$.

T.6. $\cos 2\theta = \cos^2\theta - \sin^2\theta = 2\cos^2\theta - 1 = 1 - 2\sin^2\theta$.

T.7. $\tan 2\theta = \frac{2\tan\theta}{1-\tan^2\theta}$.

T.8. $\tan 2\theta = \frac{\sin 2\theta}{\cos 2\theta}$.

T.9. $\sin^2\theta = \frac{1}{2}(1 - \cos 2\theta)$ and $\cos^2\theta = \frac{1}{2}(1 + \cos 2\theta)$.

T.10.

$$\sin(x + y) = \sin x \cos y + \cos x \sin y,$$
$$\sin(x - y) = \sin x \cos y - \cos x \sin y,$$
$$\cos(x + y) = \cos x \cos y - \sin x \sin y,$$
$$\cos(x - y) = \cos x \cos y + \sin x \sin y.$$

T.11.

$$\sin x + \sin y = 2\sin\left(\frac{x+y}{2}\right)\cos\left(\frac{x-y}{2}\right),$$
$$\sin x - \sin y = 2\cos\left(\frac{x+y}{2}\right)\sin\left(\frac{x-y}{2}\right),$$
$$\cos x + \cos y = 2\cos\left(\frac{x+y}{2}\right)\cos\left(\frac{x-y}{2}\right),$$
$$\cos x - \cos y = -2\sin\left(\frac{x+y}{2}\right)\sin\left(\frac{x-y}{2}\right).$$

T.12.

$$\sin x \sin y = \frac{1}{2}(\cos(x - y) - \cos(x + y)),$$
$$\cos x \sin y = \frac{1}{2}(\sin(x + y) - \sin(x - y)),$$
$$\cos x \cos y = \frac{1}{2}(\cos(x + y) + \cos(x + y)),$$
$$\sin x \cos y = \frac{1}{2}(\sin(x + y) - \sin(x - y)).$$

T.13.

$$\tan(x + y) = \frac{\tan x + \tan y}{1 - \tan x \tan y},$$
$$\tan(x - y) = \frac{\tan x - \tan y}{1 + \tan x \tan y}.$$

T.14. $\tan^{-1}(A) + \tan^{-1}(B) = \tan^{-1}(\frac{A+B}{1-AB})$.

T.15. $\sin 3x = 3\sin x - 4\sin^3 x$ and $\sin 4x = 4\sin x \cos x - 8\sin^3 x \cos x$.

T.16. $\cos 3x = 4\cos^3 x - 3\cos x$ and $\cos 4x = 8\cos^4 x - 8\cos^2 x + 1$.

T.17. **Law of Sines:** In any triangle $\frac{a}{\sin A} = \frac{b}{\sin B} = \frac{c}{\sin C}$. This can be extended to: $\frac{a}{\sin A} = \frac{b}{\sin B} = \frac{c}{\sin C} = 2R$, where R is the circumradius of the triangle (the radius of the circle circumscribed about the triangle).

T.18. **Law of Cosines:** In any triangle, $c^2 = a^2 + b^2 - 2ab\cos C$.

Logarithm Facts

For $a, b, c > 0$, $a, b, c \neq 1$ and for all x, we have the following facts:

L.1.

$$\log_c(a) = x, \quad \text{which is equivalent to } c^x = a,$$

$$\log_c(ab) = \log_c a + \log_c b,$$

$$\log_c(a/b) = \log_c a - \log_c b,$$

$$\log_c(a^x) = x\log_c a,$$

$$\log_a b = \frac{1}{\log_b a},$$

$$\log_b a = \frac{\log_c a}{\log_c b} \quad \text{(where } c \text{ is any base, this is the change}$$

$$\text{of base formula).}$$

L.2. $a^{\log_a x} = x$.

Sequence Facts

S.1. If the first term of an *arithmetic* sequence is a_1, the nth term is a_n, and the sum of the first n terms is S_n, then

$$a_n = a_1 + (n-1)d,$$

$$S_n = \frac{n}{2}(2a_1 + (n-1)d) = \frac{n}{2}(a_1 + a_n).$$

S.2. If the first term of a finite *geometric* sequence is a_1, the nth term is a_n, and the sum of the first n terms is S_n, then

$$a_n = a_1 r^{n-1},$$

$$S_n = \frac{a - ar^n}{1 - r}.$$

If $|r| < 1$, then the infinite geometric series, $\sum_{i=1}^{\infty} a_i$ has a sum of:

$$\frac{a}{1 - r}.$$

S.3. The sum of the first n positive integers, the sum of the squares of the first n positive integers and the sum of the cubes of the first n positive integers are respectively as follows:

$$\sum_{i=1}^{n} i = \frac{n(n+1)}{2},$$

$$\sum_{i=1}^{n} i^2 = \frac{n(n+1)(2n+1)}{6},$$

$$\sum_{i=1}^{n} i^3 = \left[\frac{n(n+1)}{2} \right]^2.$$

It is interesting to note that the sum of the squares of the first n positive integers is the sum of the squares of those integers multiplied by $\frac{2n+1}{3}$ and the sum of the cubes of those integers is the square of the sum of the integers.

Inequality Facts

I.1. For $x > 0$, the minimum value of $x + \frac{1}{x}$ is 2.

I.2. **Arithmetic–Geometric Inequality:** (AM–GM theorem):
If $a, b > 0$, then $\frac{a+b}{2} \geq \sqrt{ab}$ with equality holding if and only if $a = b$.
More generally, if $a_1, a_2, a_3, \ldots, a_n > 0$, then $\frac{a_1 + a_2 + a_3 + \cdots + a_n}{n} \geq \sqrt[n]{a_1 a_2 a_3 \cdot \ldots \cdot a_n}$ with equality holding if and only if all a_n are equal.

I.3. If $a + b$ is constant, the product ab is largest when $a = b$.

Number Theory

In what follows $a|b$ means "a divides b" and $a \equiv b \bmod m$ means $a - b$ is divisible by m (where m is a positive integer greater than 1) and $(a, b) = d$ means that d is the greatest common divisor of a and b.

N.1. If $a|b$ and $a|c$ then $a|b+c$ and $a|b-c$.

N.2. If $a|bc$ and a is prime, then $a|b$ and/or $a|c$.

N.3. If $a \equiv b \bmod m$ and $c \equiv d \bmod m$, then

$$a + c \equiv (b+d) \bmod m,$$
$$a - c \equiv (b-d) \bmod m,$$
$$ac \equiv bd \bmod m,$$
$$a^p \equiv b^p \bmod m, \quad \text{where } p \text{ is a positive integer.}$$

N.4. **Fermat's Little Theorem:** If p is a prime, and $(a,p) = 1$, then $a^{p-1} - 1$ is a multiple of p. (Said another way, $a^{p-1} \equiv 1 \bmod p$ from which it follows that $a^p \equiv a \bmod p$.)

N.5. Factor a positive integer, N, into the product of powers of primes p_k, such that:

$$N = p_1^{n_1} \cdot p_2^{n_2} \cdot p_3^{n_3} \cdot \ldots \cdot p_k^{n_k}$$

Then the number of positive divisors of $N = \prod_{i=1}^{k} p_i^{n_i}$ is

$$(n_1 + 1)(n_2 + 1)(n_3 + 1) \cdots (n_k + 1) = \prod_{i=1}^{k} (n_i + 1).$$

The sum of the positive divisors can be found by:

$$\frac{p_1^{n_1+1} - 1}{p_1 - 1} \cdot \frac{p_2^{n_2+1} - 1}{p_2 - 1} \cdot \ldots \cdot \frac{p_k^{n_k+1} - 1}{p_k - 1} = \prod_{i=1}^{k} \frac{p_i^{n_i+1} - 1}{p_i - 1}.$$

N.6. All positive integers have an even number of positive divisors except for perfect squares.

N.7. All positive integer solutions of $a^2 + b^2 = c^2$ are given by:

$$a = k(m^2 - n^2),$$
$$b = k(2mn),$$
$$c = k(m^2 + n^2),$$

where m, n, and k are arbitrary positive integers with $m > n$. If $(a, b, c) = 1$ we need $k = 1$, and m and n to have opposite parity.

Probability Facts

P.1. The number of ways of arranging n distinct things is $n!$.

P.2. If task 1 can be done in n_1 ways and task 2 in n_2 ways, and so on, and task k can be done in n_k ways, then the number of ways to perform task 1 followed by task 2, and so on, followed by task k is $n_1 \cdot n_2 \cdot n_3 \cdot \ldots \cdot n_k$. This is called the counting principle.

P.3. The number of ways to choose k objects from a set of n objects where the order does not matter is

$$_nC_k = \frac{n!}{k!(n-k)!} \cdot \; _nC_k \text{ is often written as } \binom{n}{k}.$$

P.4. The number of ways to choose k objects from a set of n objects where the order does matter is

$$_nP_k = \frac{n!}{(n-k)!}.$$

P.5. A Bernoulli experiment (also called a binomial experiment) is an experiment with the following:

    (a) There can only be two outcomes, success and failure. Call the probability of success p and the probability of failure q.

    (b) There are a definite number of trials, call it n, and an exact number of successes wanted, call it k. Then the probability of getting exactly k successes in the n trials is:

$$P(k) = \, _nC_k \cdot p^k q^{n-k}.$$

P.6. If A and B are independent events (that is, the occurrence of one in no way affects the occurrence of the other), then the probability that both events occur is given by

$$P(A \text{ and } B) = P(A) \cdot P(B).$$

P.7. If A and B are mutually exclusive events (that is, A and B cannot occur simultaneously), then the probability that one or the other occurs is given by

$$P(A \text{ or } B) = P(A) + P(B).$$

P.8. For any events A and B, $P(A \cup B) = P(A) + P(B) - P(A \cap B)$.

Miscellaneous Facts

M.1. If $[x]$ represents the greatest integer less than or equal to x, then

$$\left[\frac{x+m}{n}\right] = \left[\frac{[x]+m}{n}\right],$$

where m and n are integers, and $n > 0$, and if $[-x] = n$, then $[-x] = -n - 1$.

The ceiling function, represented by $\lceil x \rceil$, is the smallest integer, greater than or equal to x. For example, $\lceil 4.3 \rceil = 5$ and $\lceil -4.3 \rceil = -4$.

M.2. $e^{i\theta} = \text{cis}\theta = \cos\theta + i\sin\theta$.

Cross References

The cross references are intended as an easy guide to finding problems and solutions involving particular aspects of mathematics. In some cases, the problems will use more than one category, in other cases, they do not specifically use any of the listed categories.